Magnetism of Rocks

Michael Urbat

Magnetism of Rocks

Concepts, Geological, and Everyday Applications

Michael Urbat
Rhos on Sea, Wales

ISBN 978-3-662-70427-1   ISBN 978-3-662-70428-8 (eBook)
https://doi.org/10.1007/978-3-662-70428-8

Translation from the German language edition: "Magnetismus der Gesteine" by Michael Urbat, © Der/die Herausgeber bzw. der/die Autor(en), exklusiv lizenziert an Springer-Verlag GmbH, DE, ein Teil von Springer Nature 2024. Published by Springer Berlin Heidelberg. All Rights Reserved.

This book is a translation of the original German edition "Magnetismus der Gesteine" by Michael Urbat, published by Springer-Verlag GmbH, DE in 2024. The translation was done with the help of an artificial intelligence machine translation tool. A subsequent human revision was done primarily in terms of content, so that the book will read stylistically differently from a conventional translation. Springer Nature works continuously to further the development of tools for the production of books and on the related technologies to support the authors.

© The Editor(s) (if applicable) and The Author(s), under exclusive license to Springer-Verlag GmbH, DE, part of Springer Nature 2025

This work is subject to copyright. All rights are solely and exclusively licensed by the Publisher, whether the whole or part of the material is concerned, specifically the rights of translation, reprinting, reuse of illustrations, recitation, broadcasting, reproduction on microfilms or in any other physical way, and transmission or information storage and retrieval, electronic adaptation, computer software, or by similar or dissimilar methodology now known or hereafter developed.
The use of general descriptive names, registered names, trademarks, service marks, etc. in this publication does not imply, even in the absence of a specific statement, that such names are exempt from the relevant protective laws and regulations and therefore free for general use.
The publisher, the authors and the editors are safe to assume that the advice and information in this book are believed to be true and accurate at the date of publication. Neither the publisher nor the authors or the editors give a warranty, expressed or implied, with respect to the material contained herein or for any errors or omissions that may have been made. The publisher remains neutral with regard to jurisdictional claims in published maps and institutional affiliations.

This Springer imprint is published by the registered company Springer-Verlag GmbH, DE, part of Springer Nature.
The registered company address is: Heidelberger Platz 3, 14197 Berlin, Germany

If disposing of this product, please recycle the paper.

Contents

1 Introduction 1
2 The Earth's Magnetic Field from a Geological Perspective 11
   2.1 The Earth as a Dynamic System 11
   2.2 The Earth's Magnetic Field Over Time 31
3 Magnetism 57
   3.1 Different Types of Magnetism 63
   3.2 Rock Type Determines Acquisition Mechanism 68
   3.3 Size and Time 75
   3.4 Self-Reversal, Reference Systems and Arrows 80
   3.5 Magnetic Susceptibility of Rocks and Minerals 87
4 Paleomagnetism—Applications 93
   4.1 Magnetostratigraphy 98
   4.2 Moving Continents with Stored Earth Magnetic Fields 126
5 Rock Magnetism 145
   5.1 Delineation from Paleomagnetism 145
   5.2 Applications of Rock Magnetism in the Geosciences 156
6 Conclusion and Outlook 193

Introduction 1

On the way back from my wife's book club, her acquaintance approaches me, saying she has heard about my idea for a manuscript—a non-fiction book specifically for laypeople, aha!—and inquires about the topic and whether I already have a title in mind. My willing response, that it should broadly be about the Earth's magnetic field over the last 500 million years and generally about the magnetization of rocks, is met with a "You'll come up with something with an actual geoscientific background". After all, I am some kind of geologist.

After many years of dealing with a topic, whether as a hobby or professionally, some facts may seem commonly known. As in this case with (geo-)scientific research. And so one easily forgets that supposedly common terms like "Earth's magnetic field" are not part of most people's active vocabulary. This leads to situations where topics that are close to one's heart and, as in the mentioned conversation, communicated with pure enthusiasm, are perceived as preposterous fantasies. Probably just to mask an immature concept.

In the late 1990s, advertisers in London played with the aura of niche knowledge. An advertisement was placed in several lifestyle magazines to promote a spirit. In a word puzzle in the background of the advertised bottle, the terms "paleomagnetist"[1] and "smart arse" were combined (Fig. 1.1). So let's go.

Earth's magnetic field—it really does exist.

Why one would want to write a non-fiction book about something that is explicitly aimed at laypeople and not at like-minded researchers would probably have been the further topic of a conversation not prematurely ended. A book about magnetism and then about the magnetism of rocks. Magnetism, for which we humans are known to lack the senses—not to see, not to hear, not to smell, and not to taste. Something whose presence

---

[1] People who deal with the magnetism of rocks in geological history.

© The Author(s), under exclusive license to Springer-Verlag GmbH, DE, part of Springer Nature 2025
M. Urbat, *Magnetism of Rocks*, https://doi.org/10.1007/978-3-662-70428-8_1

Fig. 1.1 In reference to an advertisement in a London magazine many years ago

in everyday life does not at all fall into the category of always wanting to experience, because the Earth's magnetism will usually go unnoticed in everyday life without sensual attraction. The reference to magnetism in everyday life is even decreasing. However, sometimes it is worth taking a closer look. Perhaps to find a trace by chance, another hint, and to look further, to be surprised by what can be found. That's exactly what this book wants.

For most people, the hiking compass with its magnetic needle has given way to smartphone and GPS application. Thus, even when hiking through nature, one of the few direct references to the Earth's magnetism is increasingly fading into the background. The Earth's magnetic field works, so to speak, in the background. It performs its service rather unnoticed, but effectively. Not least by acting as a kind of shield against solar winds (cosmic radiation), thus protecting our atmosphere, among other things, maintaining a pleasant oxygen content for living beings.

This protective effect is also usually not perceived. Unless research results on the stability of the Earth's magnetic field[2] and possible collapse make their way from the

---

[2] We notice two things at this point: Magnetic fields—preferably first in the imagination of a dome-like protective shield—as well as the Earth's magnetic field have a strength and a direction (Fig. 1.2).

Strength can easily be understood with the idea of the effectiveness of a protective shield against attacking forces. Be it to protect the atmosphere against cosmic radiation or in the case of a superhero (see below) against adversaries of all kinds.

Direction is well illustrated with the use of a compass needle, which points north. Or imagine our superhero confuses the direction when setting up his magnetic force field—let's say, north pole and south pole for a bit of magnetic jargon. Perhaps with the effect it has when you twist two small bar magnets against each other and repulsion becomes attraction. More on this later.

1 Introduction

Fig. 1.2 On the importance of direction and strength of a magnetic field. In magnetism, it is the combination of strength and direction that is decisive for expected effects. This applies both to the shielding effect of the Earth's magnetic field against cosmic radiation and generally to the storage of magnetization in rocks. In the latter case, it is primarily the direction of magnetization that interests us in many applications of paleomagnetism (see polarity reversal of the Earth's magnetic field in Chap. 2)

specialist literature to the headlines of the daily press. Then warnings are issued about the consequences of an unprotected Earth, should the Earth's magnetic field collapse.

The Earth's magnetic field has indeed "flipped" countless times over the last 500 million years. We will get to know various ways to take advantage of this characteristic of the Earth's magnetic field, rather than "fearing" it. However, we do not want to speculate about the specific effects should the Earth's magnetic field collapse in this book. It should only be noted that there is no correlation between major extinction events in Earth's history and known flips of the Earth's magnetic field. And extinction events are numerous in Earth's history—far beyond the much-cited extinction of the dinosaurs at the end of the Cretaceous period about 65 million years ago.

If we want to talk about the past of the Earth's magnetic field, we cannot avoid geology. At least not without a look (or two) at various relationships in the "workings" of our Earth. This book will also provide us with such considerations.

In lectures on paleo- and rock magnetism[3], a basic chapter talks about so-called magnetic moments. In the 1990s, a perfume of the same name was once advertised by a famous tennis player. Probably such a specialized term has thus, even if unrelated and with a completely different meaning, achieved the widest distribution in general language use that one could wish for. We want to explore in this book what more is hidden behind such a term in the scientific sense and what one can do with it in everyday life.

Numbers, equations, and other beauties of mathematics and physics should be avoided in this book if possible. If only to not lose my currently only confirmed reader, my daughter, right at the starting line. Overall, the report should be technically correct, but more with a view to gaining a broader understanding of geosciences than as a collection of facts and enumeration or explanation for experts. We will get to know some examples of how this approach is to be understood already in this introductory chapter.

We can talk in general terms about a multitude of aspects of our environment and our Earth, especially in Earth's past. Perhaps to better understand the (geological) present or simply to see it with different eyes. We will talk about things that surround us every day and may seem ordinary. Or are not noticed at all, but can evoke astonishment upon closer examination. Or simply show a hidden beauty and may arouse interest.

For example, how do we know how the Earth's magnetic field behaved in the geological past, how it changed, or whether it even existed? How can we know the strength and direction of the Earth's magnetic field in Earth's past?

To the question of why such questions are asked, here is a historical note: A few hundred years ago, seafaring merchants noticed that ships navigated with a compass can land at their destination port faster. The compass promised better orientation and thus the possibility of a shorter travel route. As a result, systematic measurements of the direction of the Earth's magnetic field were recorded during the numerous sea voyages. In particular, where the compass needle points north, this was depicted in maps of the so-called magnetic declination. As a result, a kind of medieval GPS was created, perhaps even a rightful ancestor of today's satellite navigation. Including the need for updates, as the direction of the Earth's magnetic field changes not abruptly, but steadily and surprisingly quickly, and the maps had to be adjusted accordingly. We will go into the rapid temporal and spatial change of the Earth's magnetic field in more detail.

Nowadays, there are a multitude of more sophisticated ways to globally capture the current Earth's magnetic field. The corresponding measured values are recorded in maps, tables, and graphs and analyzed. But for example 100 million years ago, in the Cretaceous period, or even 4 billion years ago, in the early Earth? Obviously, we ideally

---

[3] Usually within the framework of a university study of geophysics and/or geology. The research areas of paleomagnetism (Chap. 4) and rock magnetism (Chap. 5) are taught and advanced at numerous renowned universities and research institutes worldwide. To name a few such institutions would be unfair to those not mentioned—and it would not be useful to list here well over 100 active research groups from Europe to America to Australia. Practical applications are increasingly finding their way into the industry.

need a stable recorder for this. And what could be closer in view of the mentioned time spans than to take a closer look at rocks?

Stones?

Here's the claim: Rocks store the direction and strength of the Earth's magnetic field. Depending on how a rock formed, such a recording varies in quality, but still persists over geological time spans[4]. No, as a rule, rocks are not magnets in the common sense. Nobody would think of trying to stick the collected pebbles from the river bank to the refrigerator. Nevertheless, thousands of circular, coin-sized boreholes can be found worldwide in completely ordinary rock formations. Scientists have drilled rock samples there to find out in the laboratory from these rock samples what the Earth's magnetic field looked like hundreds of millions of years ago. So are rocks magnets after all?

"It depends" would be an obvious answer to this question. Throughout this book, it becomes clear that such an answer—especially in the context of the exact natural sciences—should not be hastily interpreted as too vague. It depends on what we understand by the term "magnetic" (Chap. 3). It depends on how well we understand our geological environment and thus our environment and its interrelationships. It depends on how closely we want to look. We take a small additional look with this book. Where are our considerations to be placed?

The planet Earth is a complicated structure, subject to constant, often dramatic changes over and for millions of years. A significant aspect—generally speaking of geoscientific research—is to illuminate past states of the Earth system over the millions of years of Earth's history. Only against this background can the *current* dynamics of our planet be better assessed. Where are earthquakes more likely than elsewhere on Earth? Where should we sensibly search for resources from water to lithium? How should we evaluate present climate changes against the background of natural fluctuations in Earth's history? The list of such overarching questions is long.

No geoscientific discipline can answer such overarching questions alone. Often, less complex questions need to be answered first through research of specialists, for example: How old is a rock? How was it formed? Why was it later tectonically deformed? What happens to a rock under increasing pressure and temperature? In which direction did a river, now long dried up, flow millions of years ago? This list is also long.

We want to illuminate some examples and some background on what contribution rocks and magnetism can ultimately make to the overarching questions (Fig. 1.3).

The path leads us along the most diverse aspects of the magnetization of rocks, following the topic of the book. At suitable points, smaller snippets of geological background knowledge are always sprinkled in. Just enough, hopefully, to appreciate why, for example, dating rocks is an exciting topic for geologists and how magnetostratigraphy can help.

---

[4] We will use the term "geological time span" in the following to describe "long time". Long in geology means several million to (rather) billions of years.

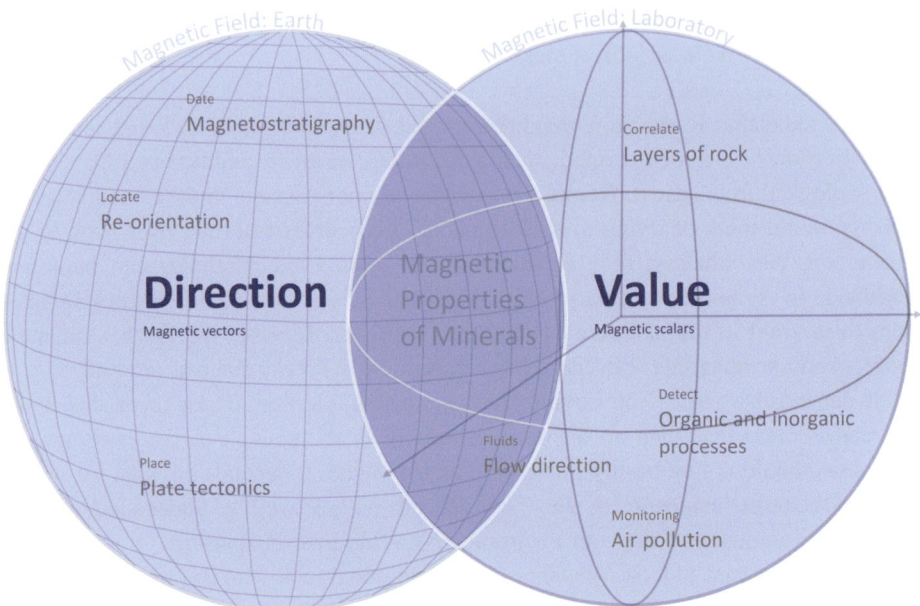

Fig. 1.3 The magnetism of rocks is hidden in the different properties of the individual rock-forming minerals. Magnetization occurs at the smallest, subatomic level and results in characteristic parameters of all minerals when examined in the laboratory (right sphere). Few of the minerals resemble the classic idea of a magnet and are already magnetized in nature by the Earth's magnetic field. In this way, directions (such as from the North Pole to the South Pole) can be determined (left sphere). Typical applications of paleo- and rock magnetism in the geosciences are within this range (subatomic to global Earth's magnetic field) as well.

Geology, or rather the geoscientific understanding of our Earth since its formation, is based on the combination of a multitude of most diverse disciplines. Geosciences use physically and mathematically shaped studies such as in crystal physics, or the technically highly specialized methods of three-dimensional seismic recordings in the depths of the Earth's crust. Geodynamic studies range from the complexly structured Earth's interior to noticeable effects of the Earth's interior on the surface such as earthquakes and volcanic eruptions. Geosciences serve to analyze past environmental conditions in the billions of years since the Earth's formation. The topic of this book—paleo- and rock magnetism—forms a puzzle piece in the search for a detailed picture of our planet. In line with the initially quoted "smart arse" Nimbus, there are indeed opportunities to consult the magnetization of rocks on all scales of dealing with the "Earth system".

The application examples discussed on the following pages are deliberately chosen to be diverse—from the most diverse areas and in interaction with different geoscientific disciplines and questions. This is to perhaps somewhat vainly emphasize the range of paleo- and rock magnetism. We use the two technical terms "paleomagnetism" and "rock magnetism" in chapter headings. Both are connected through their scientific roots, but still represent two completely different ways of utilizing magnetism in rocks.

In Chap. 4 we deal with past states of the Earth's magnetic field and what we can read out of rocks. We will see and understand, we look at vectors and only a vanishingly small number of tiny, but special minerals of the Earth's crust.

In Chap. 5 we no longer need the Earth's magnetic field to make geoscientifically relevant contributions to answering our questions. We deal with the fact that everything reacts magnetically, for example a quartz grain. It is not necessary to permanently remember everything, not even an Earth's magnetic field, to find magnetic attention.

To read this book and understand the implications of the topic, a certain geological or, more generally, geoscientific basic understanding is probably necessary. However, since such understanding should not be assumed (I don't want to scare off my daughter), we will repeatedly undertake small geological excursions at appropriate places. In view of the geological background, we cannot deviate too much from the magnetic path of this book, but just enough to not lose our footing as geoscientific laymen. Who can appreciate the descriptions of a new climbing tool if the existence of mountains and cliffs is completely unknown? But perhaps we can recognize the mountains with some thought supports, even without climbing them. And maybe even spend more time there.

In addition to geology, we will not be able to avoid thinking about how magnetization of rocks works and what magnetism actually is. But who hasn't always wanted to know what's going on at the smallest, atomic level in rocks?

As we will see, not the entire rock stores the magnetization, but only a small group of contained minerals with special magnetic properties (Chap. 3). These "magnetic" minerals are often submicroscopically small and usually only found as trace elements in rocks. For people with a corresponding research interest, however, they are consistently present in *every* rock. A quartz grain in a sandstone[5], that is, what we perceive with the naked eye as the main component of the sandstone, carries no memory of past earth magnetic fields. You have to look closer to notice the magnetically relevant grains next to the quartz grains.

The topic we are getting into also means dealing with millions and millionths, billions and billionths. We want to look at things—namely components of rocks—which are too small to see easily, and processes—namely magnetization—which have usually lasted so long or started so far back that one has to swallow three times when realizing the magnitude of these periods of time.

Billions in this context refers to years. Typical "magnetic" minerals are smaller than a millionth of a meter (or a thousandth of a human hair).

Most of us may find the pyramids with their around 4000 years very old. In geology, plus or minus 40 million years are not particularly impressive. In everyday life, some

---

[5] A rock that has formed from previously loose sand grains through the overburden of younger deposits, as the grains have wedged and baked into each other over time.

of the wealthiest people in 2022 are probably not overly excited at 40 billion (dollars in this case). Excuse the mixing of age and currency, but it's just about the overarching point of dimension. For the average earner, due to lack of experience, these sums are so high that they elude not only the imagination but also the daily relevance for covering a high standard of living in the western world. 40 billion can just about suffice to carry out a project with the aim of expanding the human horizon. Perhaps in space, the size of which is also not easy to realize. Back to the dimension of time, 40 million years (a thousandth of 40 billion!) are long enough to form a mountain range like the Alps. Much too long for the preservation of the pyramids, which would probably have weathered by then. We will make some considerations later on to better understand the immense duration of a "geological period".

As said, every stone, every rock in nature contains magnetic particles. You have to look closely, starting with a look at the respective rock itself, to be able to classify which question can be answered at all. Knowing the age of rocks is interesting in many respects, for example to reconstruct how a landscape or even the position of a continent on the globe has changed over the millions of years of earth history. Whole continents have moved many thousands of kilometers relative to each other, collided with each other and piled up mountains like the Alps or the Himalayas. Other continents have disappeared forever, been subducted. Geoscientists want to understand the earth's past in order to, ideally, better assess our current environment and its critical factors and thus also control our behavior towards nature.

"Flips" of the earth's magnetic field, in which the respective north pole becomes the south pole and vice versa, have left a kind of magnetic barcode in rocks. The duration and relative sequence of the respective configuration of the earth's magnetic field are documented for long periods of earth history in the black and white stripes. This age information can now not be seen, perhaps similar to counting tree rings in dendrochronology, with the naked eye or under the microscope, but can nevertheless be read out using scientific measurements and analyses—as always, if you know how, usually quite easily. More on this in Chap. 4.

In Chap. 5 we devote ourselves to the not so wrong claim "Everything reacts magnetically". This refers to materials that surround us every day—for example rocks, wood, plastic, even water. The claim will initially quite rightly cause most people to shake their heads and turn away rather than instill a heightened interest in further explanations on the topic due to its apparent absurdity.

So how is this to be understood: Everything is magnetic?

Back to our sandstone[6]. Let's briefly set aside the Earth's magnetic field and stones, a tiny, invisible part of which can store this magnetic field. A sandstone typically consists almost 100% of packed quartz grains. One might not generally think of associating

---

[6]Those who are unsure about such geological terms will find some explanatory hints throughout this book and hopefully be able to navigate along the written word at such points.

quartz and magnetism. However, it is a question of what we understand by magnetism. It is certain that every material, every type of mineral, shows a measurable magnetic reaction when this mineral is exposed to a magnetic field. This does not mean that such magnetization must necessarily be permanent. In fact, the magnetic reaction of most minerals fades as soon as the magnetizing field[7] is removed. In rock magnetism, quartz, like all other minerals of the Earth's crust, are therefore extremely interesting objects of study. In fact, it is precisely the "non-magnetic" minerals as the main components of the rocks that are responsible for a multitude of magnetic applications and insights. After all, these components are most likely to tell the story of the rock's formation.

Why do we want to know all this?

Magnetic properties of rocks and minerals can contribute in many ways to answer geoscientific questions. A main reason for this is that many of the minerals are very sensitive to all possible changes in their geological environment from a magnetic point of view. Or simply reflect their geological environment. This can mean that we can determine relative changes in the composition of a rock in one way or another with magnetic measurements. We will use the unusual application to reassemble a destroyed gigantic Buddha statue in the Hindu Kush and understand why the relative composition of a rock is interesting.

In a continuation of this idea, we will learn that it is worth taking a closer look at often the smallest components of a rock. Some magnetic minerals can in a way show what has happened to them during their existence. Was the temperature in the Earth's crust unusually high, have fluids flowed through the rock, or have bacteria tried to incorporate the minerals into their food chain? Some minerals reflect in their magnetic face which direction the less sensitive main components of a rock were transported from.

Sometimes magnetic investigations even provide the decisive clue to the ultimate solution of a geoscientific problem. Overarching geoscientific questions are usually of a complex nature, as they naturally deal with complex natural systems. Therefore, it is generally not to be expected that a single method, a single discipline within the geosciences, or generally a restricted perspective can lead to a satisfactory result. Paleomagnetism and rock magnetism have a contribution to make in almost all areas of the geosciences. But more on that in the course of the book.

Here is another reference to the more general perception of magnetism. It is exciting to see and read how Magneto[8] evolves from the supposed villain to the superhero. This usually happens powerfully and loudly with the help of various magnetic abilities, for example the use of magnetic force fields. But in this book, as mentioned, we want to deal more with the inconspicuous and silent effects of magnetism in rather everyday situations. After all, there are more silent superheroes than one might think.

So let's start imagining a journey through the world of Earth's magnetism.

---

[7] Such a field is often caused by a magnetic measuring device during a corresponding examination. But of course, this also applies to the Earth's magnetic field.

[8] A protagonist from the X-Men of the Marvel Comics series. Since the 1960s.

The Earth's Magnetic Field from a Geological Perspective

2.1 The Earth as a Dynamic System

An important approach when reading this book will be to view Earth's crust from a geological perspective as a dynamic, constantly changing, moving, destroying, and creating garment of this Earth. Rocks have been forming, changing, deforming, and were often being destroyed again in cycles for billions of years. Even though these usually take millions of years.

With this approach, the entire Earth's crust can be divided into a few larger fragments (tectonic plates), which are constantly moving relative to each other. They collide, deform, and are piled up into mountains along the colliding plate boundaries. All mountain ranges that may come to mind have essentially formed in this way, albeit at different phases of Earth's history. It is no coincidence that the highest mountains, such as Mount Everest, are located in an ongoing geological collision between India and Asia.

Mountains, on the other hand, where the collision has been completed, are eroded over time, i.e., gradually worn away by wind and weather. The German low mountain ranges were significantly higher in the Earth's history than they are since we humans admire them. Today, only the rounded bases of these mountains remain, so to speak. The comparison holds true for the principle of mountain formation and the subsequent inevitable decay. Not all mountains once looked the same, for example, reached the same heights and number of mountain ranges. The determining factors[1] are too varied in a worldwide and geological comparison.

Without briefly making ourselves aware of the structure of the Earth and, in this context, particularly the significance and dynamics of the Earth's crust, we cannot proceed

---

[1] These include the type and quality of the piled-up rocks as well as the intensity, angle, period, and type of plate collisions.

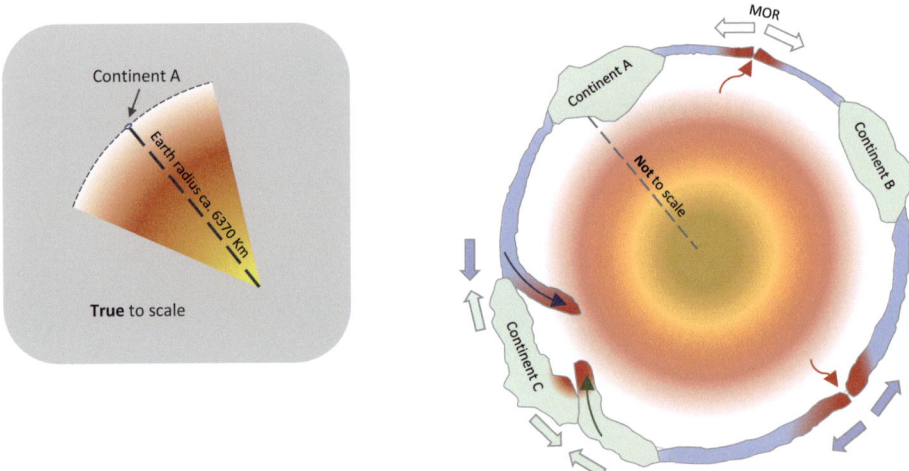

Fig. 2.1 Highly schematic and non-scale cross-section of the Earth. To engage with the topic of "magnetization of rocks," it helps to understand the Earth's crust as a dynamic, constantly changing garment of our Earth. On the left, a realistic proportion of the thickness of the Earth's crust to the Earth's radius is shown. This allows us to better classify the constant movement within this thin Earth's crust. In particular, the illustration can serve us as a rough framework for the cycle of rocks. Rocks are formed, for example, at mid-ocean ridges (MOR; diverging arrows on the right), are folded into mountains (converging arrows), or are melted again (during their subduction). In the course of the chapter, we will go into more detail on the significance of dynamics for the magnetization of the Earth's crust, but also on the significance of the vaguely indicated layered structure of the Earth for the generation of the Earth's magnetic field (see also Fig. 2.12)

at this point.[2] Without knowledge of earlier processes of the dynamic development of the Earth's crust, many areas of human resource exploration (water, ores, minerals, hydrocarbons, etc.) would be left entirely to chance. Knowledge of similar geological configurations in Earth's history helps enormously to search in the right place and to understand what to expect where and what not.

From the geosciences, we learn that the Earth, in a first approximation, has a layer-like structure consisting of crust, mantle, and core. The so-called Earth's crust—the cooled surface on which we walk—has about the thickness of an eggshell compared to the diameter of the Earth (alternatively also peach or apple peel; Fig. 2.1). Not a

---

[2] In other words: The safe location of a new garden shed can probably be determined by a brief local inspection of the condition of the subsoil and without further knowledge of geology. A skyscraper, on the other hand, requires, in the truest sense of the word, a deeper consideration of geology to ensure its stability. Otherwise, it might be built on sand. And we want to aim high with our new insights into the magnetism of rocks.

2.1 The Earth as a Dynamic System

particularly uniform shell, though. There are significant differences in thickness[3], composition[4] and stability. As said, all parts of the crust are continuously moving.

All of this is highly simplified; other factors play a role. For example, it is important how the colliding plates are structured. Geologically, oceanic and continental crust are distinguished, which are each made up of different types of rocks with different physical properties and differ in their thicknesses by about an order of magnitude. Kilometer versus tens of kilometers thickness.

The perhaps hastily drawn comparison from everyday life: Heavy truck collides with light small car. Truck hits truck etc. At what angle, at what speed does the collision of the rather similar or dissimilar partners occur? Possible combinations as well as the respective environmental conditions[5] are as diverse as the respective results. The comparison with a car collision is strongly flawed—essentially, in the geological scenario, it is not a spontaneous energy conversion, but a long, persistent, but immensely powerful pushing against each other.

But also in the Earth's crust, the scenario will be significantly different depending on whether two continents collide, a continent meets oceanic crust, or oceanic crust of different ages and therefore different density collide. Plates at different locations on Earth move very slowly, only a few centimeters per year, yet the Earth's crust moves at some places twice or three times as fast as at other places. There are many reasons for this. For example, those sitting on the outside of the carousel cover more distance than those near the axis of rotation. The speed as a ratio of distance per time (km/h) is higher, even if the angular speed[6] is the same. This comparison also is somewhat lame with regards to Earth's crust and its individual plates. These move in first approximation on the spherical surface of the Earth around so-called Euler poles. This is the axis of the carousel[7]. Nevertheless, the image of a rather everyday traffic collision may help to understand the multitude of factors involved in a plate collision, especially with regard to the present book, in order to understand the multitude of possible deformations of a formerly horizontally deposited and magnetized (!) rock layer due to plate tectonic movements.

Often a part of the crust is pushed into the depths during the collision of two plates and subsequently melted. In another place, new crust is formed by cooling and solidifying when magma rises from the mantle. This happens primarily along hundreds of kilometers long so-called mid-ocean ridges (MOR) in the depths of the oceans. As we will discuss in more detail later in the chapter, a MOR is thus created from the fault line

---

[3] In geology, this is called "thickness" when it comes to rock sequences.

[4] Continental versus oceanic crust and their respective typical rock (see also Chap. 3).

[5] Wet road, icy conditions, dry intersection, curve, highway etc.

[6] How fast the carousel is rotated.

[7] The Indian continent has been colliding unusually fast and directly with Asia starting a little more than 50 million years ago. The Himalayas and neighboring mountain ranges up to Pakistan and Afghanistan are the result.

along two breaking plates. If the Earth's sphere is not to expand or shrink over millions of years, a certain balance between newly formed (MOR), disappearing (subduction zones) and folded (mountains) Earth's crust must be maintained. If one area moves, all other crust parts must follow.

So all four corners of our garden are constantly in motion. Of course very slowly. Very, very slowly, but if I move only 1 mm further each week, I still make progress, and after 100 million years (!) I would have already walked from Cologne to the equator. As I said, for a geologist, 100 million years is not even a particularly impressive long period of time. Counted backwards from today, this just takes us to the middle of the Cretaceous period, i.e. the youngest Mesozoic era. Since the formation of the Earth, one could have covered much further distances.

The oldest rocks on Earth are over four billion years old, although due to the aforementioned movements and recycling tendencies of the Earth's crust, they are not found too often anymore. Usually, the age of the Earth is noted as 4.6 billion years. In our example with a speed of 0.5 cm per year, we could circle half the Earth at the equator in 4 billion years (approx. 20,000 km). A few million years are already enough time to send continents around the world, fold up mountains, erode mountains again and carry the erosion debris further and further down the valley, over rivers to the sea. There, these deposited sediments are then piled up again into a mountain range, folded and sheared in a possible next collision of the Earth's plates.

Another example (Fig. 2.2): If one undertakes a journey from the North Pole to Cologne Cathedral and somewhat nonchalantly *equates time with distance*,[8] the Cambrian[9] begins in the North Sea at about the height of Edinburgh. The Mesozoic begins roughly at the Dutch coast, the Cretaceous ends at the German-Dutch border. The first modern human (Holocene) encounters us about 12 m in front of the cathedral portal.

Depending on how, when, in what order, and under what climate and environmental conditions rocks were formed in the geological past, or possibly later altered by very high temperatures and pressures, rock layers are of different interest to us humans today. The mentioned factors in rock formation play an important role, for example, in the formation of resources and their possible use. This also includes the possibilities of groundwater storage and its flow behavior—or the nature of the subsoil for new buildings, tunnels, or roads. Sometimes the interest is purely for the sake of basic geological research. As said, one can say very little about the current state of the Earth if the past is not known and at least understood in principle.

---

[8] Just a thought experiment.

[9] See especially the following explanations of the geological time scale compared to the geomagnetic time scale in Sect. 2.2.

2.1 The Earth as a Dynamic System

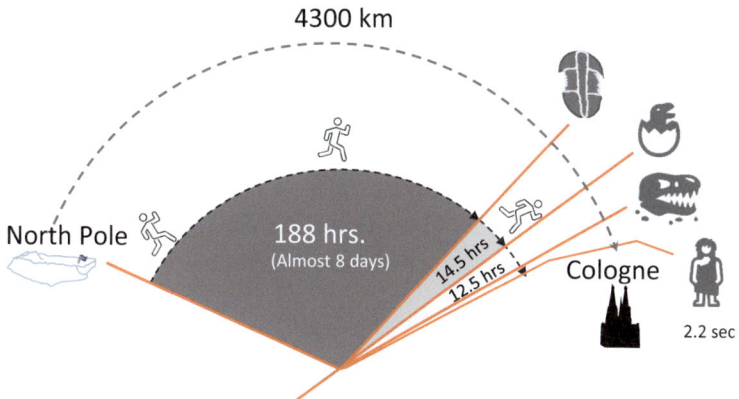

Fig. 2.2 In this book, we also talk about extreme dimensions in terms of time and size. The illustration shows one of the many possible analogies to realize the geological age of the Earth. Here, a time span of 4300 million years is compared to the duration of an "ultramarathon" from the North Pole to Cologne (approx. 4300 km). For about eight days we ran through lifeless wasteland (Precambrian) until we encountered the first primitive life forms at the beginning of the Cambrian. The ninth day of the journey included all periods of the development of flora and fauna in Earth's history, including the appearance (Triassic) and extinction (Cretaceous) of dinosaurs. The first human (*Homo sapiens*) only met us seconds before arriving at the cathedral portal. Most geological research, including the rock magnetic studies at the center of this book, refer to the last 12 hours of the journey. However, for the conception of geological time spans, the absolute value of 12 hours is less important[10] than the relative duration of time compared to the total duration of the journey of about nine days

Another geological classification should be recalled here: the so-called rock cycle. This is not another hint at how lively our rock world is, but the very pragmatic summary of the three (!) main types[11] of rocks on this Earth:

1. Igneous rocks: Solidified from a hot magma.
2. Sedimentary rocks: Existing rock weathers and is eroded, the components are transported, redeposited, and solidified.
3. Metamorphic rocks: Igneous or sedimentary rocks are exposed to high temperatures and/or pressures in the Earth's crust, thereby altered, usually fundamentally.

A rock always consists of individual components typical for this particular rock (minerals, let's say a bit casually, grains). Depending on whether the rock originally solidified from a magma, these could be such different rocks as basalt or granite. Anyone who has perhaps once visually compared the typical blue-black cobblestones of basalt in an old

---

[10] Resulting from an arbitrary assumption of a running speed of 20 km/h and a distance of 4300 km.

[11] There are justifiably many (very many) different rock names for different rocks. Granite, basalt, sandstone, or slate may be named spontaneously. However, all these rocks can be assigned to one of the three main categories mentioned in the text.

town alley with a granite kitchen countertop will immediately agree. The components (minerals) are fundamentally different, but it will immediately be noticeable that a basalt is so fine-grained that practically no individual grains can be seen without a magnifying glass. In the context of this book, we cannot go into the details of the differentiation[12] of igneous intrusive and extrusive rocks. As a short example, let us consider the difference whether a magma cools slowly at a shallow depth below the Earth's surface and forms large well-crystallized components, or the same magma is explosively hurled into the air from a volcano and cools very quickly. In the latter case, individual components remain small and poorly crystallized.

When a so-called sediment, composed of already eroded and transported components (grains), is deposited and over time solidifies back into a rock, the components are usually easily recognizable to the naked eye. Sometimes a good microscope is needed to distinguish and identify the components. Sometimes there are only a few different components, sometimes many different minerals, that make up a sediment. Individual components (minerals, grains) are always present.

Interestingly, the grains of a particular rock do not all have exactly the same size, but there is something like an average size depending on the respective formation process. This statement is valid for all three main types of rocks. In the case of sediments, for example, quartz is relatively hard compared to other minerals and therefore remains the longest after the erosion of a rock. Quartz grains take a long time to assume a more and more similar shape and size through breaking and rounding. Some grains deviate in size above or below the average prevailing grain size. In the simplest case, this results in a bell curve[13], if all grains of a rock sample were measured, counted, and listed (e.g., Fig. 3.9).

So let's note: Rocks, regardless of their "provenance", typically consist of minerals, the type, composition, and size of which depend on the respective formation conditions of the respective rock. More importantly: Every rock *forms* at a specific point in Earth's history. The duration of its formation compared to its usually millions of years long geological existence is short. After the formation of the rock, most rocks weather more or less slowly, sometimes being completely destroyed again.

---

[12] Another one of these geological terms: During the cooling of hot magma, different minerals crystallize according to certain laws in a certain order. This continuously changes the composition of the remaining magma, as the components of the formed minerals are removed from the overall mixture.

[13] More details in Chap. 3, especially since we will also learn there that minerals with special magnetic properties typically have a similar grain size distribution. And the magnetic "abilities" of a mineral change dramatically with its grain size.

Good for our magnetic topic: All rocks, no matter how different they are, always contain the finest "magnetic" minerals[14]. Most of these minerals are much smaller than the other components of the respective rock. Magnetic minerals are on the order of much less than 1/1000 mm, so micrometers! For comparison: In medical terms, air pollutant particles are categorized from $PM_{2.5}$ to PM_{10}. This fine dust is so dangerous because the particles are so small that they can penetrate into the finest branches of the lung. On the other hand, they are not visible to the naked eye, unless in a dust cloud. $PM_{2.5}$ stands for "**p**articulate **m**atter" with a size of 2.5 micrometers. An average human hair is about 30 times as thick. Magnetic particles in rocks, namely those that can store the Earth's magnetic field (Chap. 3, 4 and 5), are about 1/100 smaller than $PM_{2.5}$.

These illustrious[15] magnetic particles form in many geological scenarios at the same time as the rest of the respective rock (Chap. 3), but can often be altered more easily[16] than the surrounding (we understand: dynamic, constantly changing) rock.

After all this mental engagement with the constant changes of our seemingly stoic rock world (if we disregard earthquakes and other disasters for a moment), this brings us to stable features of our Earth: North and South Pole[17].

The pole is a familiar term when considering the Earth. A chain of associations might include polar expeditions, dog sleds, snow boots, and solitary flags in white desolation, marking the spot of the North or South Pole. The, let's be more specific, *geographical* pole is thus commonly understood as a defined, supposedly unchangeable location at the northernmost or southernmost point on the globe. In this view, the Earth's axis of rotation emerges at the pole, and the longitudes converge there.

Upon closer inspection, it becomes clear that this supposedly fixed point, in the case of the North Pole in the "eternal" ice on the surface of the frozen ocean, is actually occupied by ever-changing, drifting ice floes (Fig. 2.3). A flag of success, left to its own devices, will over the years slowly but steadily be carried further and further away on the drifting ice floes—until the distance and current position of the flag even raise doubts as to whether the former goal was ever reached. Or until the ice melts.

Immovable—anchored deep on the ocean floor—a titanium flag has marked the location of the (imaginary) axis of rotation and thus the North Pole for several years. Upon closer inspection, the Earth's crust also moves relative to the axis of rotation. We will

---

[14] The term "magnetic minerals" is used somewhat loosely in this book mainly for a small, illustrious group of minerals with the special ability to store the magnetization of the Earth's magnetic field (see especially Sect. 3.1). This usually does not mean that one should imagine these minerals in the classic sense as magnets.

[15] We will learn in the following chapters that these particles are not only special, small, and rare, but also rare in relation to the other components of a rock.

[16] For the storage of an Earth's magnetic field, this initially seems bad (Chap. 3). In Chap. 5 we will learn about the advantages of sensitive magnetic minerals.

[17] Here initially meant geographically, but with a sure allusion to the following classifications of the Earth's magnetic field.

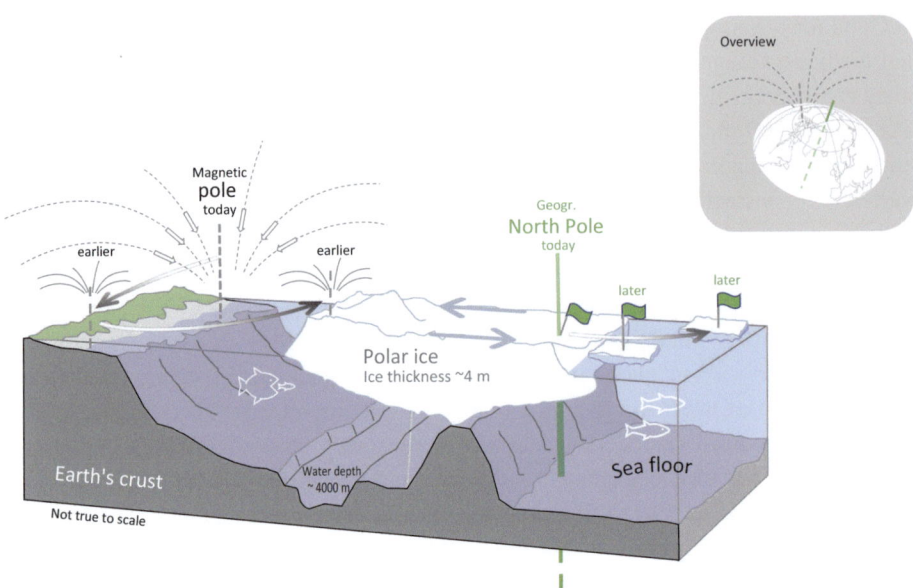

Fig. 2.3 Highly schematic representation to improve the understanding of the different poles on the northern hemisphere. The geographical pole remains stationary if you don't look too closely. The magnetic pole is constantly moving

delve into this dynamic of the Earth's crust several times in this book, if only to better understand the magnetization of these rocks.

Plate tectonics (Sect. 4.2) and the associated movement of the Earth's crust relative to the axis of rotation are disregarded when imagining the flag, correctly placed for eternity, to mark the geographical pole. This becomes clearer when we imagine the South Pole on the Arctic continental plate, which is only covered by ice. The flag moves in relation to the axis of rotation, even if it is only movements in the millimeter/centimeter range corresponding to the speed of continental movement.

Despite this more meticulous view when defining the geographical pole, we would probably have no qualms about fixing our flag in a suitable place in the ice of the Arctic or Antarctic on the timescale of a human life and understanding it as a permanently marked point.

Many views and definitions, here regarding our considerations of the geosciences, are a matter of the reference system. We will note several times in this book that results and facts can only be understood in relation to a specific reference system. Just as the movement of the Earth's crust plates seems insignificant within the framework of a human life, but over millions of years it folds up a mountain. A reference system can refer to a temporal framework, but also to a spatial determination or a combination of both. In particular, we need at this point already a hint of the importance of reference systems to engage in the following explanations of the symmetry of the Earth's magnetic field.

2.1 The Earth as a Dynamic System

Is at least the Earth's axis of rotation immovable, even if individual layers (here ice floes and the rock crust) move relative to each other and relative to the axis? No, Seth Carlo Chandler[18] and other researchers show that the Earth's axis of rotation is not temporally and spatially fixed. At least if we want to be accurate to about 10 m and about ten years. The Earth wobbles in its rotation around its own axis. The resulting so-called (geographical) pole wander is roughly spiral-shaped and drifts overall in a westerly direction.

Over geological time scales, this seemingly small movement of the axis is viewed differently: Similar to the example given at the beginning, the flag would have moved a distance of 100 km after 10 million years with a floe movement of only 1 cm per year. And, if not already mentioned, 10 million years are rather a short period in geology.

To not lose sight of everything, we should highlight one essential point. In geological considerations, and in relation to this book, considerations about the Earth's magnetic field, the choice of a suitable time frame and the coordinate system[19] is of crucial importance. We will learn about the advantages of a uniform reference system in Chap. 3. First, we will create a bit more confusion with an overview of the Earth's magnetic field as a global "object".

With the *magnetic* pole of the Earth, it is much more complicated at first glance than for the, as said, relatively easy to determine geographical North Pole[20]. A magnetometer would indicate a point several hundred kilometers away from the (geographical) North Pole if we were to search for this magnetic pole now, today. However, our hiking compass[21] should actually point north and thus to the geographical North Pole. Moreover, just a few years ago, we would have determined this magnetic point several kilometers away at a different location than today. And this is regardless of the movements of ice floes, the Earth's crust, or even the rotation axis itself. The magnetic pole moves (Fig. 2.4). Constantly. And even at a noticeable speed by human standards and in human time spans. The good news is—similar to the general understanding of the geographical

---

[18] If you think of the axis of rotation as a rod that is stuck through the Earth at the poles, you would see a certain imbalance in the rotation. "Chandler wobble" is a term for this. For example: Mandelbrot, B.B. and McCamy, K., 1970. On the Secular Pole Motion and the Chandler Wobble. *Geophysical Journal International*, Volume 21, Issue 2, November 1970, Pages 217–232, https://doi.org/10.1111/j.1365-246X.1970.tb01777.x.

[19] The familiar geographical longitudes and latitudes of the Earth provide us with a certain coordinate system. We can say: We are heading west because north and south have been defined as *fixed* reference points on Earth.

[20] In this book, we repeatedly use specific terms as examples and mainly for simplicity, here "North Pole". With equal justification, "South Pole" could also be written, especially since both poles are symmetrically and fixedly positioned relative to each other.

[21] In a compass, a freely swinging magnetic needle shows us where north is. This needle rotates due to the Earth's magnetic field and not because of the Earth's rotation axis or our definition of longitudes and latitudes.

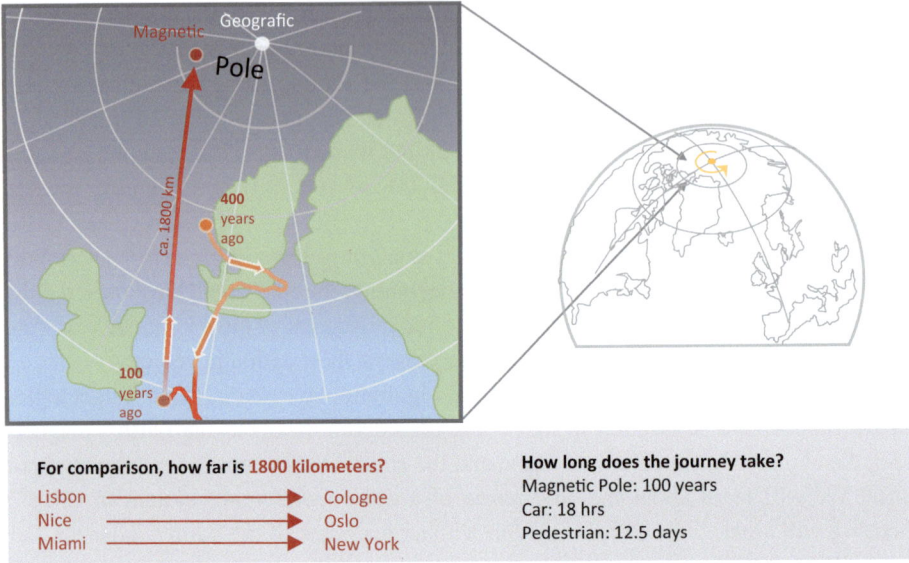

For comparison, how far is 1800 kilometers?		How long does the journey take?
Lisbon	→ Cologne	Magnetic Pole: 100 years
Nice	→ Oslo	Car: 18 hrs
Miami	→ New York	Pedestrian: 12.5 days

Fig. 2.4 This is a highly schematic representation of how the *Magnetic* Pole moves relative to the *Geografic* Pole and thus to the Earth's rotation axis. We may understand the movement of the pole here as a symbol for the movement of the entire Earth's magnetic field. This movement is irregular and rarely follows a nearly straight line, as shown. Overall and over longer periods of time, the magnetic pole performs a meandering dance in all directions around the Earth's axis. A period of 400 years is indicated, which is only a small part of the "circumnavigation" of the geographical pole. For better understanding, the realistic movement of the magnetic pole of 1800 km in a period of 100 years is indicated. Location examples illustrate what a distance of 1800 km means.

pole—a simplification is also possible for the Earth's magnetic poles, which allows us to conveniently work with the otherwise highly variable Earth's magnetic field in time and space: In paleomagnetism, the so-called geocentric axial dipole hypothesis (GAD) is used. More on this follows.

Before we engage in hypotheses, we should first get a somewhat better understanding of this mysterious magnetic field of the Earth. A key question is the relationship and shape of the Earth's magnetic field in relation to the Earth's sphere or, in other words, to the commonly used cardinal directions and the up and down in relation to the Earth's surface. Ideally, we want to understand the Earth's magnetic field as a global structure with a simple relationship to the longitudes and latitudes of the geographical system, as if the Earth's magnetic field had a shape similar to the grid of the known longitudes and latitudes. This would make it possible, for example, to infer a specific position on Earth from the magnetization in a rock.[22] Similar to a compass, whose needle points north and additionally allows the latitude of the location to be read from the respective inclination

---

[22] Don't worry, there's a whole chapter (Chap. 4) coming up on this.

angle of the needle relative to the horizontal. For everyday use, this sounds cumbersome and very antiquated in times of GPS. But we are looking for the Earth's magnetic field in the geological past and are slowly working our way towards the millions of years old directions of the Earth's magnetic field stored in rocks. So, patience.

In a further step of approximation, we will visualize the shape of the Earth's magnetic field, the so-called field lines, in relation to the Earth's sphere. For this, we consider the field of a bar magnet, perhaps still known from school (Fig. 2.5). A vivid conception of the shape of the Earth's magnetic field then follows in Fig. 2.6.

From the consideration of the rapidly moving magnetic poles (Fig. 2.4), we already suspect it. If I were to take a suitable measuring device—a magnetometer—at this moment, step outside and measure the field lines of the Earth's magnetic field, then from all these measurements I would measure a very complex magnetic structure. Suppose I take these measurements simultaneously worldwide. This structure of field lines would hardly correlate with the Earth's longitude and latitude. With a bit of good will and above all slightly squinted eyes when looking at these measurements, one might guess something like the shape of the magnetic field lines as in Fig. 2.6 and 2.7. But that's

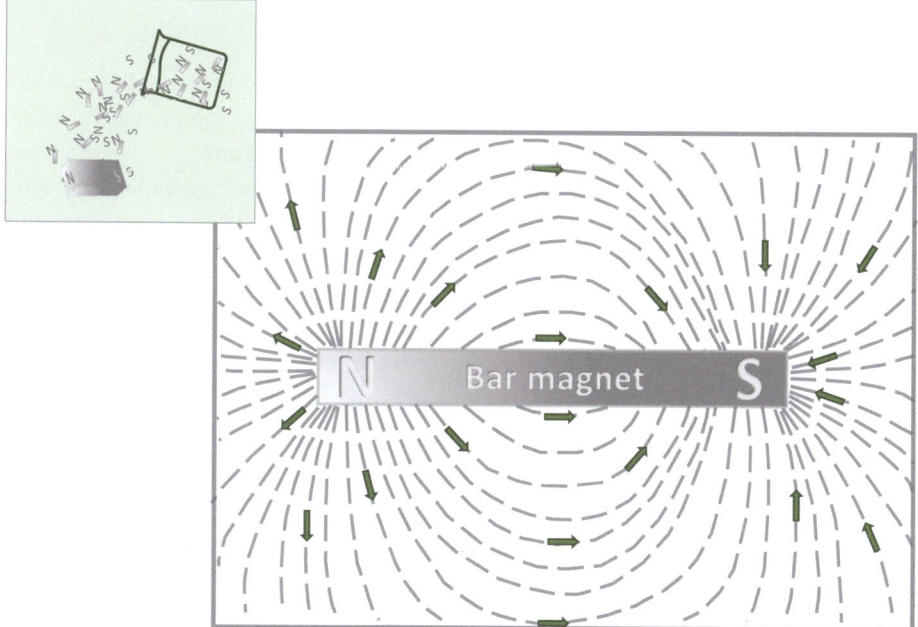

Fig. 2.5 Humans hardly perceive weak magnetic fields, such as the Earth's magnetic field. With the help of a simple experiment, in which small iron filings or simply small magnets are poured onto a larger bar magnet, we can make the so-called field lines and thus, in a certain sense, a magnetic field itself visible. The filings align in a characteristic way along the field lines of the bar magnet. The experiment is particularly interesting because the Earth's magnetic field has a very similar shape and thus forms the basis for numerous considerations in this book

Fig. 2.6 Let's imagine planting a huge bar magnet into the Earth. This is what the field lines of the Earth's magnetic field could look like. In the further course, we will spend some thoughts on why the Earth's magnetic field cannot be generated by a huge bar magnet in the Earth's interior. In particular, we will learn why the shape of the Earth's magnetic field on the Earth's surface can still be compared with the field lines from Fig. 2.5

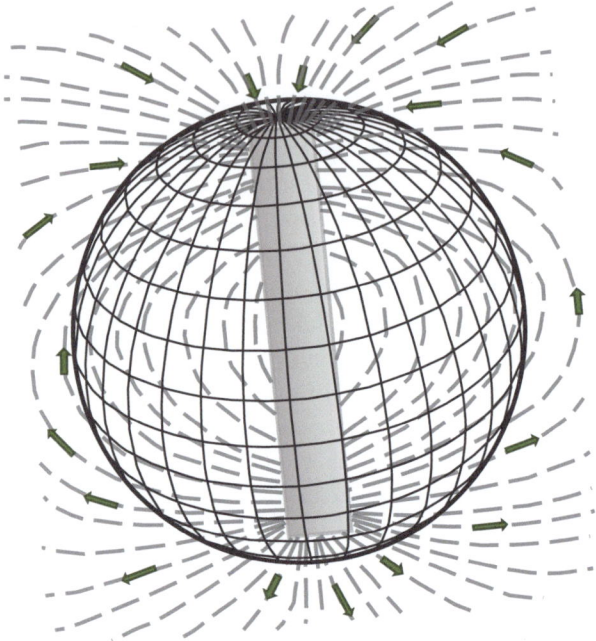

about it. Not a single field line will show a straight course, and not even the magnetic north and south poles are diametrically opposed, but at most, approximately, in the expected region on the northern and southern hemisphere of the Earth.

At first, this does not sound very promising. For paleomagnetic applications including dating by magnetostratigraphy, the Earth's magnetic field would ideally be as symmetrically structured as possible, as shown in Fig. 2.7.

The Earth's magnetic field would ideally be simply correlated with the geographical reference system of longitude and latitude at every point on Earth. Ideally, it would be possible at every location on the Earth's surface to predict the direction of the field lines entering or exiting the surface. If we look at Fig. 2.7, a magnet needle hanging on a thread would align itself parallel to Earth's sphere along the field lines at the equator and point north like a compass needle[23] (Fig. 3.2). At the pole, the tip of the needle points steeply downwards directly towards the pole. From the equator to the pole, from low to high latitudes, the angle of the needle to the Earth's surface becomes increasingly and continuously steeper. In other words, it would thus be possible to deduce the geographical latitude from the inclination of the needle to the Earth's surface. A conclusion about the geographical latitude alone would not be particularly remarkable and can be achieved more easily in other ways, without resorting to the Earth's magnetic field. However, we

---

[23] Physically speaking, the needle points to a south pole.

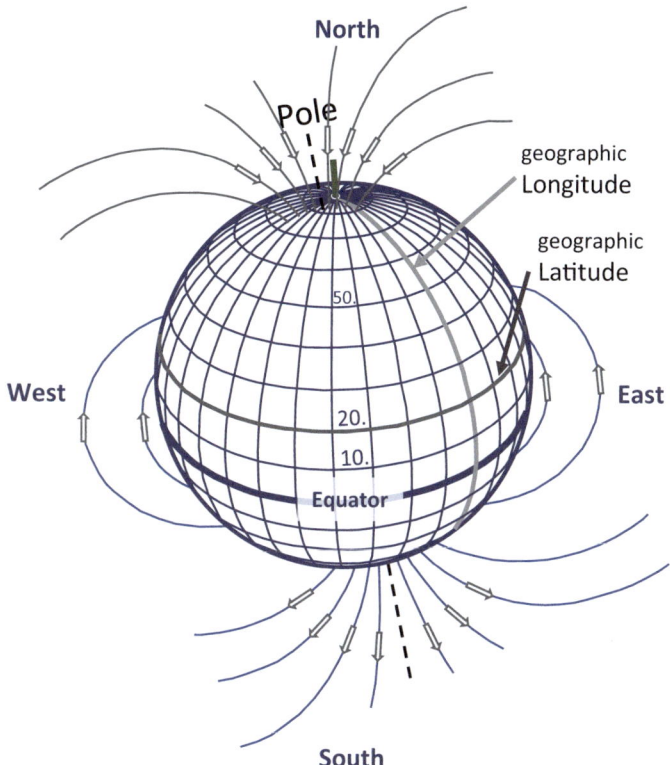

Fig. 2.7 One way to specify positions on Earth is based on dividing the surface into geographical longitude and latitude. Longitudes run at equal intervals from pole to pole and are "numbered". Latitudes run "across" them, in the same way as the equator, and are also numbered. 90 latitudes from the equator to the North Pole and also 90 latitudes from the equator to the South Pole. If we compare the Earth's magnetic field with the quite symmetrical (dipole) field of a bar magnet, a direct relationship between geographical longitudes and the course of the magnetic field lines can be established. A snapshot of the field lines of the Earth's magnetic field unfortunately only roughly meets this requirement. In addition, the field is highly variable over time, and there is no bar magnet inside the Earth. Further down, it is explained what further steps are necessary to nevertheless achieve the desired correlation of longitudes and field lines for long periods of Earth's history

will encounter and appreciate the principle of a simple correlation of latitudes to the shape of the Earth's magnetic field as a basis for success for many of the magnetic applications we will learn about in this book.

Now, as said, the current Earth's magnetic field is much more complex. It is not symmetrical and certainly not a dipole in its entirety—and thus not comparable to a bar magnet with a north pole and an opposite south pole.

To better understand the shape of the Earth's magnetic field at the Earth's surface, it helps to illuminate how the Earth's magnetic field is generated: Of course, there can be no bar magnet in the Earth's interior. Apart from the knowledge of geoscientists about the so-called layer structure of the Earth, it would be much too hot in the center of the Earth with several thousand degrees Celsius. Iron would be molten, but more importantly: Looking at physics, we recognize the necessity for a certain order at the smallest, atomic level of a permanent magnet. High temperatures, heat in the physical sense, mean the supply of energy, and this energy will cause too much movement at the subatomic level to be able to maintain the order necessary for permanent magnetization. It's as if someone were blowing into a nicely arranged ensemble of loose petals.

And even at moderate temperatures in the Earth's interior, the effect of a permanent magnet would have significantly decreased after over 4 billion years since the Earth's formation. Permanence in the sense of an ordered structure is also only a matter of time in nature, or rather, a matter of probability, of ending up in a certain arrangement out of trillions of possible arrangements[24] if no external force helps. In the same sense, somewhat more generously interpreted, order means energy, and energy often decreases over time. And with it the ordered state.

We have already used petals in our somewhat plakative analogies, another analogy would be the idea of always sitting in a perfect, ergonomically ideal position at the desk while writing. I rarely manage to do this in the long run.

Following the second law of thermodynamics, entropy always increases. In other words, a state of disorder is more likely as long as there is no external specification for order. A magnetizing field, for example, tries to provide such a specification. The intensity of a "permanent" magnetization[25] then decays exponentially again, even at ordinary room temperature.

Here is the opportunity to remind ourselves of another important factor in geology: time. The decay of a magnetization takes millions of years at the Earth's surface. Especially when studying rocks, there is usually enough time. When we delve deeper into the matter in Chap. 3, almost in the truest sense of the word, we will get to know the term "relaxation time".

The generation of the Earth's magnetic field can best be compared with a dynamo process, not with a permanent bar magnet (see below under "Earth's magnetic field originates in Earth's interior"). The example of the bar magnet is primarily used because of the external field lines. The actual external field, the actual resulting field lines of the Earth, especially in the Earth's interior, in the area of the Earth's mantle, rather resembles a pot of spaghetti that is constantly being stirred, than the nicely ordered field in our experiment with the iron filings. What would be identified as a north or south pole is rather the north pole or the south pole than another direction. Several smaller "secondary poles" are also in play.

---

[24] Ordered state in the sense of aligned magnets.

[25] Let's assume, a bar magnet, left to itself, i.e., without further influence of an external magnetizing field.

2.1 The Earth as a Dynamic System

In the common scientific view, there are actually two different types of magnetic poles (see below under "Magnetic Poles—a Definition Question"). Here, however, the *definition* of such an Earth magnetic pole plays the decisive role. What is the respective pole to be used for? Are questions to be answered that are based on the daily change of the magnetic field, or is a more stable position of a pole averaged over a longer period of time required, for example?

Consequently, the (however defined) magnetic north pole does not initially coincide with the geographical north pole. The Earth's magnetic field is not correlated with the Earth's axis of rotation. As a result, a pole of the Earth's magnetic field or these different types of magnetic poles will be found at different and temporally changing geographical locations on Earth.

The geocentric axial dipole hypothesis (GAD) underlies paleomagnetic applications. A hypothesis that equates the Earth's magnetic field to the symmetrical field of a magnetic dipole, that is, the field of a bar magnet, whose longitudinal axis also lies parallel to the Earth's axis of rotation. How is this to be understood and especially to be accepted as meaningful? The following explanations should provide us with answers.

The axial dipole hypothesis
Surely one or the other will ask why the Earth's magnetic field can be described as an apparently rigid, albeit global, three-dimensional structure, when supposedly the magnetic poles move quite quickly (Fig. 2.4). Wasn't it initially claimed that the Earth's magnetic field as a whole is in constant change and anything but a symmetrical structure? Not even a dipole it is (Fig. 2.5). The short answer is: yes and no. The somewhat longer answer means briefly dealing with the advantages and possibilities of temporal and spatial averaging.

Assuming a sheep can move freely in a pasture, and the animal walks back and forth across the entire field without stopping (Fig. 2.8). At any given time, one could specify the current position of the sheep in the pasture. A position is given, for example, in relation to the boundaries/sides of the field. But any other reference system would be equally conceivable. An instantaneous position of the sheep obviously has nothing to do with the center of the field. After only 10 minutes, the sheep has not been everywhere in the pasture, it has mainly grazed the western area. If one wanted to give a representative position of the sheep in the last 10 minutes, one would probably intuitively choose the center of this western area. Not the center of the entire pasture.

After 1 hour, the sheep has been equally often in every region and in every corner[26] of the pasture. Perhaps it has never actually touched the exact center of the pasture, but probably one would, again intuitively, call this center of the pasture the sheep's place of stay in the last hour. In other words: 1) Depending on how fast the sheep moves and how large the pasture is, there is a period of time that is sufficient to visit the entire pasture. 2)

---

[26] Figuratively, it should be a circular pasture as in Fig. 2.8.

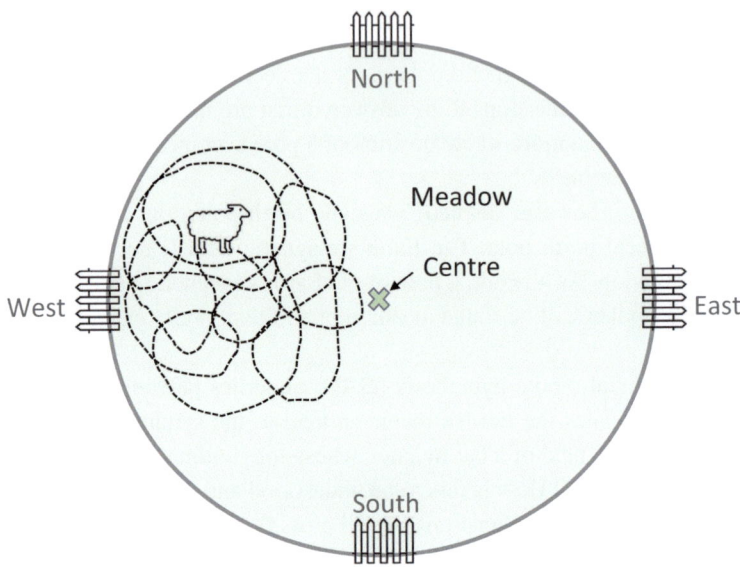

Fig. 2.8 Where is a sheep located on a pasture on average over a certain period of time?

For this period (here arbitrarily chosen 1 hour), the center of the pasture then appears to be a reasonable point to indicate the location of the sheep.

If one were to determine the average location of the sheep not intuitively, but mathematically correctly, the calculated point would correspond to the intuitively assumed point, **if** the sheep has actually been equally often and equally long at every point in the pasture. If the animal has made an unnoticed lap of honor at one point, the calculated center would still be calculated near the center of the pasture, but not exactly in the center[27].

The sheep actually brings us to the geocentric axial dipole hypothesis (GAD)[28]. One could now mock the apparent comparison of the magnetic poles with wandering sheep or even point out the obvious weaknesses of such an analogy. However, we only want to understand three basic pillars for the establishment of the geocentric axial dipole hypothesis in paleomagnetism. For simplicity, let's use only the intersection point of a magnetic

---

[27] The lap of honor pulls the average location of the sheep—so to speak—slightly out of the center of the pasture towards the area of the lap of honor. The greater the deviation from the path, better said the preferred stay in a region of the pasture, the greater the deviation from the center of the pasture will be.

[28] GAD: thus an Earth's magnetic field, which is hypothetically generated by a bar magnet exactly in the center and parallel to the Earth's axis of rotation.

pole on Earth (Fig. 2.3) instead of the entire three-dimensional structure of the Earth's magnetic field:

- Averaged over a sufficiently long period of time, the magnetic pole (the sheep) corresponds to the intersection point of the Earth's axis of rotation (the center of the pasture; Fig. 2.8).
- The sufficient period of time for the EMF is on the order of several thousand years (not much in geological terms), compared to the proposed hour for the sheep in the pasture.
- A few centimeters of sediment layer may have accumulated over a period of several thousand years. A typical centimeter-sized rock sample "contains" so to speak the movements of an earth magnetic pole over several thousand years. A magnetic measurement of a corresponding rock sample would yield a lump sum value for this accumulation period (of a few thousand years).[29]

In other words, it is appropriate to view the Earth's magnetic field as a structure similar to the network of geographical longitudes and latitudes we are familiar with. Under one condition: not the current field of the Earth is considered, but an Earth's magnetic field averaged over several thousand years. This is practical for the study of the Earth's magnetic field in the Earth's past, especially since a high temporal resolution is not conducive to many geoscientific questions.

To be a bit more precise: In fact, only about 95% of the (current) Earth's magnetic field can be described by a magnetic dipole (bar magnet). This applies to the current magnetic field of the Earth with all its irregularities. The irregularities are due to the remaining 5% of the field and are referred to as the non-dipole field. Since the current Earth's magnetic field is irregular, the Earth's magnetic field cannot be described equally well as a dipole everywhere on Earth. At some places and at some times, the non-dipole component may be significantly above 10%. This leads us to three different definitions of magnetic poles on Earth. Only one of them coincides with the geographical pole (North and South Pole).

Magnetic Poles—a matter of definition

The current exit point of the Earth's magnetic field—at which a tilt of the field lines against the Earth's surface of 90° is currently measurable—is referred to as the magnetic pole (see points A_{North} and A_{South} in Fig. 2.9). The dipole field of a bar magnet describes the approximately 95% of the current field of the Earth, if one does not think of the bar magnet exactly parallel to the Earth's axis of rotation, but tilted by about 11° against the axis. Points B_{North} and B_{South} in Fig. 2.9 show the exit points of the longitudinal axis of such a bar magnet, whose dipole field is symmetrical, but not in line with the

---

[29] The rate at which sediment builds up varies greatly depending on the depositional environment; a geological rock address is therefore part of paleomagnetic investigations. More on this later.

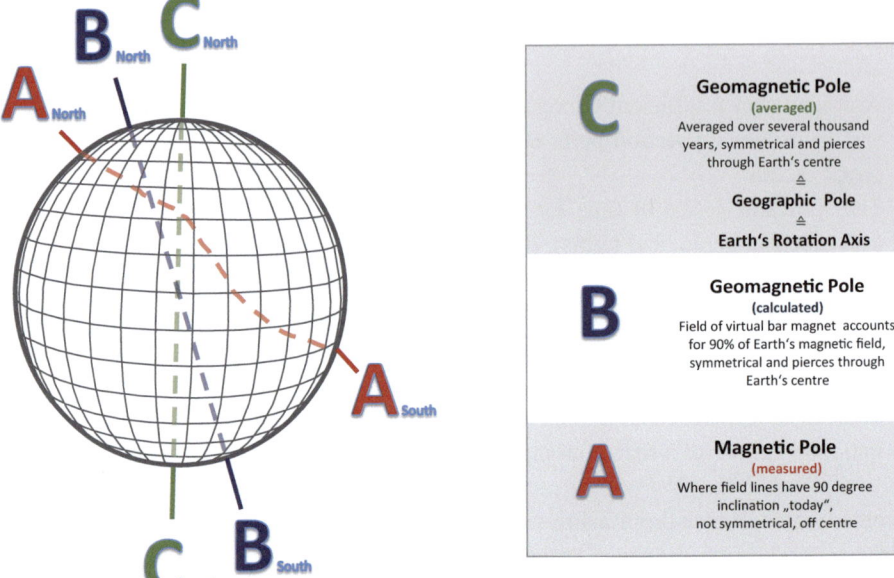

Fig. 2.9 Time is an important factor when working with the Earth's magnetic field in the geological past. Only when averaged over several thousand years does the geomagnetic pole theoretically lie on the Earth's axis of rotation and thus exhibit a symmetry that can be worked with well in paleomagnetism under the assumption of the GAD. This apparent sleight of hand can fortunately be substantiated with actual measurement data and verifiable calculations. The hypothesis holds up, especially since most magnetic rock samples represent a period of several thousand years. For example, this is the time it took for a certain "amount" of sedimentary rock to form. Magnetic measurements in this sense usually represent a temporally averaged Earth's magnetic field

geographical longitudes and latitudes. The points B_{North} and B_{South} are referred to as geomagnetic poles. If one were to not only tilt the bar magnet, but also move it a few kilometers out of the center of the Earth, the result of the modeling would improve slightly again. Since the points B_{North} and B_{South} represent the symmetrical part of the current Earth's magnetic field, it is not surprising that the geomagnetic poles, like the magnetic poles, continuously change their position over time.

The positions of the "point" B in Fig. 2.9 and similarly determined geomagnetic poles of earlier states of the Earth's magnetic field, averaged over a period of several thousand years, are compatible with a bar magnet (i.e., dipole, sheep). The poles C_{North} and C_{South} (Fig. 2.9) now coincide exactly with the Earth's axis of rotation (center of the pasture, geographical pole). A comparison of the respective movements of the different poles is offered by Fig. 2.10.

Earth's magnetic field originates in Earth's interior
As already mentioned before, there is no bar magnet in the Earth's interior. However, the dipole component of the Earth's magnetic field, generated in the Earth's interior in a

2.1 The Earth as a Dynamic System 29

Fig. 2.10 Comparison of pole movements in addition to Fig. 2.9. Schematic. The irregular movements of A lead to B. The movements of B lead to C

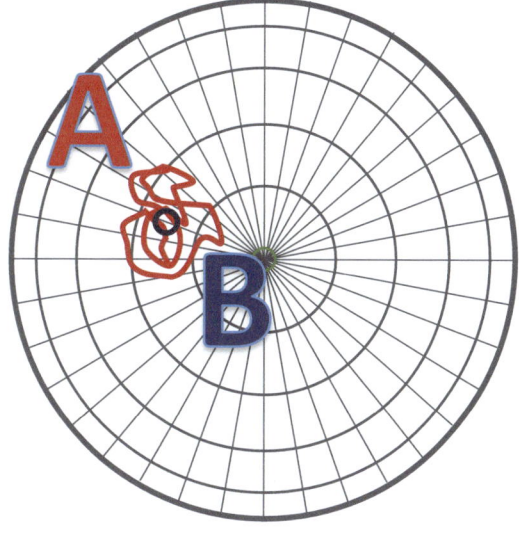

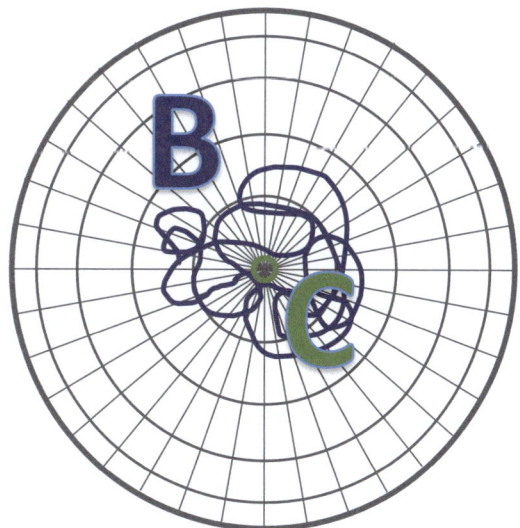

complex dynamo process, can be described by similar field lines, which are also associated with a bar magnet. Hence the simplified image of a tilted bar magnet in the Earth's interior. In Chapter 1, we have already briefly touched on the actual causes for the existence of the Earth's magnetic field. So much should be added to this topic, especially to get a better idea of why the Earth's magnetic field is constantly undergoing change. Changes, which in extreme cases lead to the irregular polarity reversals of the magnetic poles in Earth's history.

In this chapter, we use the term "dynamo process", often also called geodynamo, for the generation of the Earth's magnetic field. To be able to generate a magnetic field for an entire planet in this way, essentially three components are needed:

1. An electrically conductive (for example, iron-containing "thick") liquid (for an entire planet, a whole lot of it)
2. Convection, where the liquid is set in motion by the interaction of heterogeneity[30] and pressure and temperature differences (warm air at the radiator rises)
3. Rotation of the whole, which gives the (upward) movement an additional directional component (often spiral-shaped due to rotation)

We have also already referred to the layered structure of the Earth, especially with regard to the thin Earth's crust in relation to the Earth's radius. The Earth's core, as the name suggests, occupies the central area inside the Earth. The center of the Earth is at a depth of over 6000 km. The so-called inner Earth's core is solid and consists mainly of iron and about 20% nickel. This inner core has a radius of about 1200 km. Over a narrow transition zone, the inner core transitions into the outer Earth's core, which is liquid and also consists mainly of iron and nickel. This outer belt of the Earth's core has a thickness of about 2250 km.[31]

The temperatures in the inner Earth's core are around 5000 °C, from the center outwards and further through the outer Earth's core it becomes increasingly cooler. Relatively at least, because it remains thousands of degrees Celsius hot. The pressure inside the Earth is also significantly higher than in the Earth's crust, for example. (Put differently: A mass that weighs on our shoulders would push about five to six times harder inside the Earth). The liquid material of the outer Earth's core moves according to the physical principle of convection—in the direction away from the Earth's core. So hot material rises. Don't worry (about emerging cavities inside the Earth!): When the material has reached the outer boundary of the Earth's core, it cools down again due to the lower temperatures there, becomes heavier (denser) and sinks back into the depths. The convection cycle begins anew.[32]

So we have an electrically conductive liquid (1st component, see above) made of iron and nickel inside the Earth as well as further prerequisites for convection (2nd component). Now we are still missing the kinetic energy of a rotation (3rd component). And

---

[30] As a metaphor, one might think of blueberries in an otherwise homogeneous yogurt as an example of heterogeneity.

[31] To better visualize: If the Earth were a small gymnastic ball, the *entire Earth's core* would be the size of a football. The *inner Earth's core* would correspond to the size of a tennis ball.

[32] Note: This convection in the Earth*core* should not be confused with the convection currents in the Earth*mantle,* which would be discussed in connection with the plate tectonic movements in the Earth*crust*.

that is the Earth's rotation[33]. Due to the Earth's rotation, the rising conductive liquid is guided into spiral paths—magnetization is induced. The spiral movements in the outer Earth's core run parallel to the Earth's rotation axis, as would be expected for the described magnetic field lines on the Earth's surface.

In fact, the Earth does not rotate uniformly when viewed in its layers, but at different speeds due to the different properties from layer to layer[34]. The inner Earth's core may even rotate in the opposite direction to the rest of the Earth at times. What remains to be noted in the context of this book is that the generation and maintenance of the Earth's magnetic field over millions of years are the result of complex physical processes inside the Earth.

This makes it at least easier to understand why the Earth's magnetic field is subject to constant fluctuations. The field reversals also occur so irregularly in Earth's history, as a flip results from the—almost one is tempted to say: random—complex interplay of many properties of the Earth's interior. By properties, we mean the constantly mutually influencing physical parameters, on the far from complete list of temperature and pressure, composition and viscosity (thickness in common parlance) of the liquid[35], rotation and convection, etc. In fact, there is no regular temporal pattern throughout Earth's history that the field reversals follow. Sometimes it takes millions of years, sometimes only a few thousand years, in particular, no sequence of equally long or similar long intervals is anywhere repeated in the geological record. In Chap. 4 we will see how this can be used in paleomagnetism.

2.2 The Earth's Magnetic Field Over Time

A magnetic barcode is created

Observations and considerations about magnetism and the magnetic properties of rocks have been made by humans for over 2000 years. At that time, especially in China. Over 750 years ago, the scholar Pierre Pèlerin de Maricourt[36] wrote down his findings on the magnetic dipole properties of magnets, among other things. Petrus Peregrinus refers to physical experiments he conducted himself. Following his, briefly

---

[33] Coriolis force.

[34] Somewhat like a ball in the center of a rotating bowl full of water does not necessarily rotate at the same speed as the surrounding water due to its inertia.

[35] A liquid in the physical sense does not necessarily have to be comparable to water. Even a rather thick porridge may be referred to as a liquid, especially we must be open to new ideas in our associations with the term "liquid" in view of the combination of extreme temperatures and pressures inside the Earth, for example with regard to flowability. Whether the liquid is more like water or more like porridge.

[36] Petrus Peregrinus in Latin.

titled *Epistula de Magnete*, "scientific letter" to an acquaintance, one of the first (the first of) scientific books ever is particularly worth mentioning: Over 300, also scientifically dark years, marked by pandemics, wars and scientific doubt, had passed when William Gilbert wrote his fundamental book on various topics of magnetism and geomagnetism. Among other things, Gilbert postulates—we rightly recognize—that the whole earth is a large magnet.[37] In 1839, the mathematician and geodesist Carl Friedrich Gauss published his *General Theory of Geomagnetism*. In particular, there are calculations on the geomagnetic poles. The realization of reversals of the Earth's magnetic field from measurements on (volcanic) rocks first appeared at the beginning of the 20th century, especially between 1903 and 1906 by Bernard Brunhes[38]. His work was corroborated over 20 years later by independent measurements also on volcanic rocks by Motonori Matuyama[39]. In recognition, the prevailing configuration of the Earth's magnetic field for the last 780,000 years in paleomagnetism is referred to as the Brunhes Chron. The time of a reversed field configuration before that as the Matuyama Chron. From the realization of individual reversals, a nearly complete record of the many hundreds (!) of reversals of the Earth's magnetic field in geological history was created in the following years. We cannot honor all contributions within the scope of this book, but we will try to understand the principle of creating and using this black-and-white barcode. We more correctly refer to the barcode as the geomagnetic polarity timescale. But first, we need to make some considerations about geological timescales themselves.

The geological timescale—a classification
As for the geological age of rocks of the Earth's crust, different scientific stratigraphic methodologies are integrated *in the* geological timescale and, thus, tied to the absolute chronological sequence of Earth's history. More correctly formulated, the geological timescale is constantly updated according to the latest state of research. It is ratified by the International Stratigraphic Commission (IUGS); the underlying scientific work is based on the research of countless institutions and individuals worldwide. Essentially, two, albeit closely related, efforts can be mentioned: On the one hand, research groups are dedicated to specific periods of Earth's history, on the other hand, research in all periods of Earth's history is advanced in respective research directions—for example in paleomagnetism, paleontology, isotope chronology etc.[40]

---

[37] W. Gilbert, 1600, *De Magnete, Magneticisque Corporibus, et de Magno Magnete Tellure*.

[38] For example, 1906, Research on the direction of magnetization of volcanic rocks, *J. Phys. Théor. Appl., 4ème série, 5,* 705–724, two more publications in 1905.

[39] Matuyama, M., 1929. On the direction of magnetisation of basalt in Japan, Tyosen and Manchuria. *Proceedings of the Imperial Academy of Japan,* 5: 203.

[40] A summary of this comprehensive work is provided, for example, by the current, two-volume edition *A Geologic Timescale 2020* by Gradstein et al. (eds.), Elsevier, 1390 pages.

2.2 The Earth's Magnetic Field Over Time

An example to illustrate the complexity of a geological age scale: Anyone who delves into the geological time scale will notice the hierarchical order of the named time periods. For example, terms that may be more familiar to those outside the field, such as Cretaceous, Jurassic, or Devonian, are only found in third level in the hierarchy under the register Period/System. Cretaceous and Jurassic together belong to the overarching era of the Mesozoic (so to speak the *Middle Ages of the Earth* for the period since the development of the first life on Earth; Fig. 2.11), while Devonian belongs to the Paleozoic era *(Ancient Earth)*. Both Cretaceous and Jurassic, as well as Devonian, are each subdivided into different numbers of *epochs*/series. The Late Cretaceous (Upper Cretaceous) is in turn subdivided into the stages Maastrichtian, *Campanian*, Santonian, Coniacian, Turonian, and Cenomanian as the fifth level in the hierarchy. If the subdivision of the geological time scale may seem overly fine scaled at first glance, it should be noted: The *Campanian*, for example, covers a period of a full 11.5 million years of Earth's history.

Each of the mentioned demarcations of all stages, epochs, periods, and eras is scientifically justified, for example, by the first appearance or extinction of certain life forms, striking environmental developments in Earth's history, natural disasters, or any combinations of such events. Therefore, a radiometric, absolute age must be determined for all boundary lines. Take the date of 66 million years for the much-discussed C-T boundary between **C**retaceous and Paleogene (as the older section of the period originally referred to as **T**ertiary).

In everyday life, this boundary is better known as the extinction of the dinosaurs, usually without reference to the many other fundamental changes in life and living conditions on Earth. But at what location on Earth and exactly at which point in a sequence of rock layers is the point to be marked, below which the rocks are older than—as strikingly selected in our example—the "extinction of the dinosaurs"?

Generally, a selected type locality, which best represents a change in Earth's history, is referred to as a Global Stratotype Section and Point (GSSP). But—to mention just a few criteria for the extensive scientific investigations behind the geological time scale—is the timing of a change exactly the same everywhere on Earth, or do striking geological events have a different local expression in other parts of the Earth? How, for example, can a GSSP be transferred to other rocks at other locations on Earth?

The respective state of scientific knowledge leads to the decision about a boundary line. New findings—also in the geosciences, new and more precise measuring devices and laboratory methods, algorithms, and computer technologies for modeling and analyses are continuously being developed—may lead to a revision or refinement of existing boundaries, even though fundamental changes to the timescale have become increasingly rare in recent decades of research.

To elaborate on our above example a bit: It took many years of research and scientific discussions to be able to link and prove the extinction of the dinosaurs with the impact

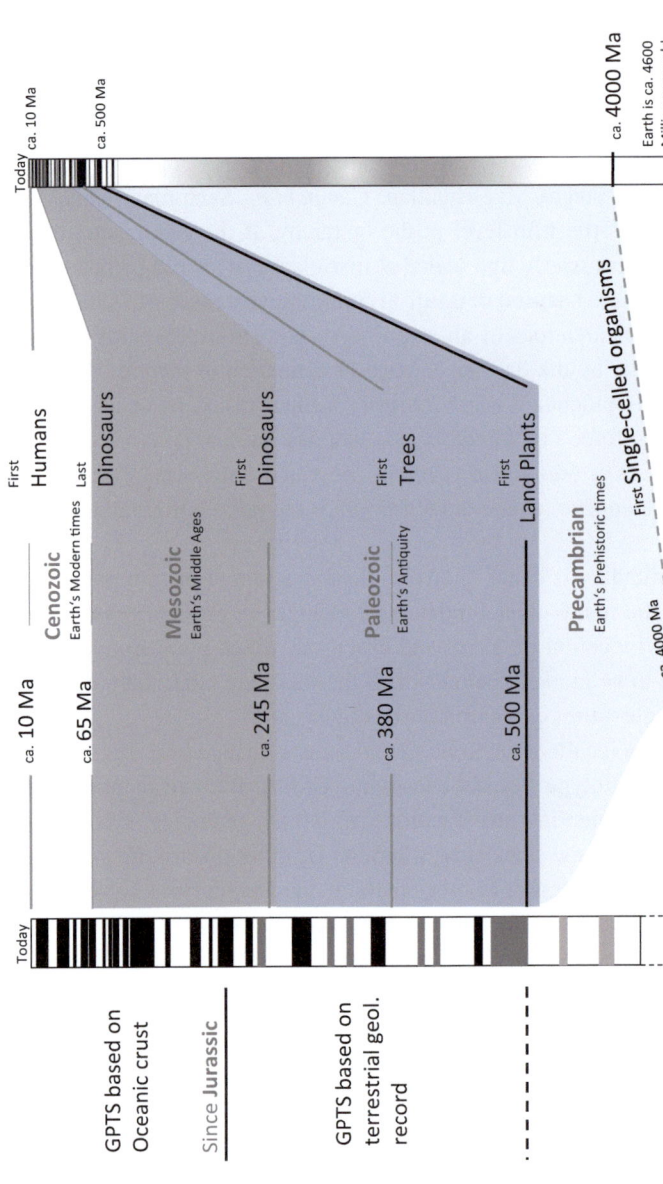

Fig. 2.11 The geomagnetic polarity time scale (GPTS) records all confirmed polarity reversals of the Earth's magnetic field in Earth's history. In particular, the last 200 million years are known in detail. The left column shows a greatly simplified, schematic GPTS (see Gradstein et al. (eds.), *A Geologic Timescale 2020* for a detailed representation of the current state of knowledge) and, highlighting selected key data of Earth's history, compares it with a likewise schematic, but to scale time column of the last 4 billion years of Earth's history. The still existing uncertainty in the older sections of Earth's history is illustrated by the partly gray areas of the "barcode". Records of the Earth's magnetic field from the oldest Earth history with ages of over 500 million years are rare. This is due to the also only sporadic preservation of Precambrian rocks, their often strong metamorphic overprinting or the poor recording of the Precambrian Earth's magnetic field from the outset. The environment of the Precambrian, as the time before the first plants and animals, differed significantly from our current environment. From a magnetic point of view, for example, frequent and widespread lightning strikes on a barren Earth's surface were directly responsible for the deletion of any records of the Earth's magnetic field in the rocks

of a meteorite[41] on Earth in present-day Yucatán. The impact ejected enormous amounts of rock material from the impact zone. The high pressure and temperatures of the impact significantly altered the rocks mineralogically and deposited them locally (hundreds of kilometers in radius). Finer particles (ashes) were carried high into the atmosphere and globally distributed by wind systems and later, possibly after years, deposited again. The resulting tsunami waves from the impact left different traces in the sediments at the end of the Cretaceous period, individually at different places on Earth and, as in the previously mentioned examples, with different effects in the highlands than on the ocean floor. Flora and fauna (including dinosaurs) experienced different, albeit perhaps only temporally delayed, effects of the impact depending on the species, habitat, and proximity to the impact site of the meteorite. Nevertheless, many different aspects of the Earth's habitat must be correlated and coordinated for the purpose of boundary determination and integration into the geological timescale. New findings from further geoscientific investigations will thus continuously refine the date of a boundary determination.

Ideally, the complete geological timescale includes a globally usable, consistent correlation from a multitude of chronostratigraphic markers at all time levels. These include:

- Radiometric age including error margin: As a rule, the accuracy of age determination decreases with increasing age. Depending on the age of the rocks to be dated, different methods are used for absolute dating, for example, based on the uranium-lead decay series, argon or carbon isotopes.[42]
- Integrated chronostratigraphy including magnetostratigraphy and biostratigraphy.
- Sea levels: They change globally depending on climatic (for example, more or less water bound as ice) as well as plate tectonic changes in the configuration of continental and oceanic crust, whereby different amounts of space are available for the same amount of sea water. Sea levels are an integral part of the analysis of the dynamic environmental development of the Earth.
- Plate tectonic phases closely linked to changing sea levels: Due to the continuous movements of the Earth's plates, there are repeated phases in the course of Earth's history, for example, when all continents meet in one supercontinent (*Pangaea*). This "convergence" is then characterized by prominent mountain building, until the convergent movement tendency is reversed into a breaking apart and moving away (divergent).

---

[41] Approximately 10 km in diameter, e.g.: Smit J., Hertogen J., 1980, An extraterrestrial event at the Cretaceous–Tertiary boundary. *Nature 285,* 198; Schulte, P. et al., 2010, The Chicxulub Asteroid Impact and Mass Extinction at the Cretaceous-Paleogene Boundary, Science, Vol 327, Issue 5970, pp. 1214–1218.

[42] Note, not every rock necessarily contains the appropriate minerals on which such radiometric dating can be performed. Possibly, the decay process of suitable minerals has been disturbed since the deposition of the rock, for example, by plate tectonic processes and associated high pressures and temperatures in the Earth's crust. In other words, it is fundamentally not possible to pick up a rock and a measuring device gives us an absolute age without further ado.

The Geomagnetic Polarity Timescale
The principle of dating with the phases of different directions of the Earth's magnetic field is simple. First, we will create a complete picture—a standard scale, so to speak—from worldwide measurements on rocks, which shows when the Earth's magnetic field was polarized how in the Earth's history (Fig. 2.11). So if we want to date an unknown rock, we measure its magnetization direction and determine the age by comparing it with the standard scale. That sounds simple. However, we should dedicate a few more pages of this book to the topic. Even the path to creating a standard scale of the Earth's historical pole flips, the geomagnetic polarity timescale, or GPTS (TS comes from timescale), is long.

The complete recording of the constantly changing states of the Earth's magnetic field is far from complete even after decades of paleomagnetic research. Even though most periods of the last approximately 500 million years are quite well documented, there are still entire periods, such as the Carboniferous, in which a lot is known about the Earth's magnetic field, but just as many questions remain open. Millions of years remain only patchily documented. One reason lies in the nature of the typical rocks from such sections of Earth's history (weak magnetizability, strong weathering, change in plate tectonic phases, etc.) or at least their patchy preservation (more on this later).

We mark periods of normal polarization black, phases of inverse polarization white. Black indicates an Earth's magnetic field with a polarity that corresponds to the current field and is therefore referred to as normal in paleomagnetism. White illustrates periods in Earth's history when the North and South poles were reversed. In paleomagnetism, this state is referred to as reversed or reverse polarity. Exact criteria are necessary to read the globally prevailing polarity of the Earth's magnetic field from rocks with sufficient certainty. A type locality, dating, independent confirmation at other locations on Earth, many samples, and especially exact laboratory methods and quality criteria are the minimum requirements.

If all phases were of equal length, a uniform (equidistant) black-and-white stripe pattern would result. This does not sound very helpful for dating—after all, every period in Earth's history would have the same magnetic pattern.

In reality, normal or reverse phases in Earth's history now have a very individual duration—from a few thousand years to millions of years. The resulting black-and-white pattern thus resembles a conventional, albeit significantly longer barcode on the goods when shopping. This makes the magnetic pattern much more interesting for dating rocks, because ideally dating is done by comparing patterns: long white, followed by short black, followed by even shorter black, etc., or in any other conceivable combination. This individual pattern in a rock sequence is then compared with the GPTS and the age is read off for the corresponding pattern. So while a change from black to white or vice versa occurs with every polarity change in Earth's history, the pattern from several consecutive polarity changes in combination with the individual durations of these polarities reflects a specific section in Earth's history. Similar to a fingerprint.

2.2 The Earth's Magnetic Field Over Time

Magnetic Stripes in the Ocean Floor

The Earth's crust in the area of the oceans has reliably recorded the polarity changes of the Earth's magnetic field at least in the younger (!) Earth's history—let's say, the last 200 million years until about the end of the Jurassic period. As we will see, the ocean floor offers an ideal recorder of the different states of the Earth's magnetic field for several reasons, not least because the way the oceanic crust is formed allows a fairly undisturbed, continuous recording. On "land", i.e., in the area of continental Earth's crust, this is much more complicated. Even before the Jurassic period, the oceanic crust recorded the Earth's magnetic field, but this older crust has already been subducted again in all oceans due to plate tectonic movements and melted again in the depths of the Earth's mantle (Fig. 2.12). The magnetic information was erased during the melting process when the oceanic crust subducted at the subduction zones[43]. Sometimes oceanic crust is not subducted, but pushed onto another (continental) plate and thus preserved. However, such crust, referred to as ophiolite, is not ideally suited to preserving magnetic

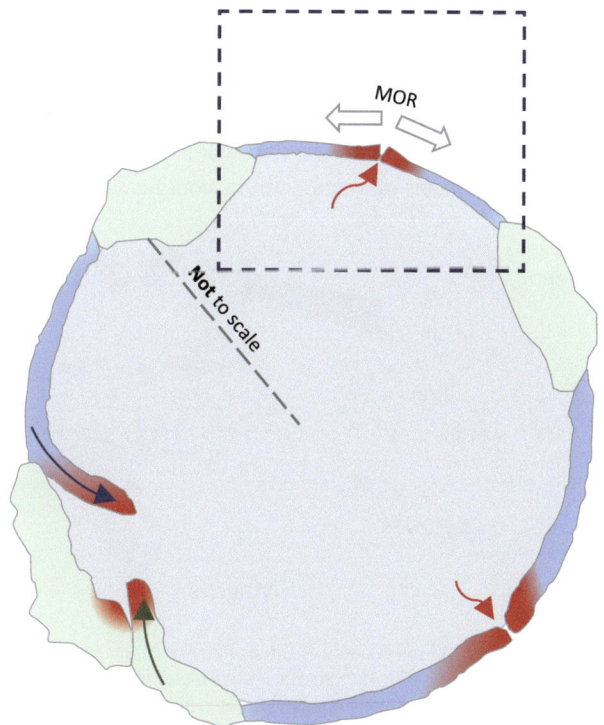

Fig. 2.12 Oceanic crust is primarily formed at mid-ocean ridges (MOR). The illustration reminds us, in reference to Fig. 2.1, that Earth's crust is also destroyed at other locations, so-called subduction zones. With a few exceptions, the Earth's oceanic crust is geologically young, older crust has already disappeared again. Just as about 2/3 of the Earth's surface is covered by water, oceanic crust also makes up the largest surface area of the Earth's crust

---

[43] The average age of oceanic crust is around 64 million years, or about the end of the Cretaceous. With the exception of an unusual preservation of older crust in the Mediterranean, there is hardly any crust older than the Jurassic (200 million years) worldwide. For example, Seton, M. et al. (2020). A Global Data Set of Present-Day Oceanic Crustal Age and Seafloor Spreading Parameters. *Geochemistry, Geophysics, Geosystems,* Volume 21, Issue 10, AGU.

information about a previous Earth's magnetic field, even if the crust may date from a time before the Jurassic period, due to the high pressures and temperatures prevailing during the collision.

At this point, we recall the image of the constantly changing Earth's crust. New Earth's crust is formed in the oceanic realm by the intrusion and subsequent cooling of hot lava from the Earth's mantle—that is, from a few kilometers deep inside the Earth from the area directly below the Earth's crust. Upon cooling, this lava records the direction (polarity) of the then prevailing Earth's magnetic field (Fig. 2.13). As a keyword from later chapters in this book, the recording is done according to the principle of heat and cooling (Chap. 3) of a thermoremanent magnetization. The most important magnetic mineral is (titano-) magnetite in basaltic lava. The ocean floor, of the Atlantic for example, thus becomes a gigantic barcode for paleomagnetism.

A pattern of this (imaginary) black and white stripes symbolizing the polarity of the Earth's magnetic field runs approximately parallel to the coasts, for example of Africa and America, and symmetrically on both sides of the MOR. Fig. 2.14 outlines the principle of such a magnetically striped ocean floor. In reality, however, there is a significantly higher number of polarity changes in the crust of the Atlantic, for example between America and Africa (or any other ocean), than the illustration shows. Strictly speaking, there are as many polarity changes as there are between today and the Jurassic period[44].

A MOR is explained in a few words as follows: A continent breaks apart over hundreds of thousands of years, with its continental crust being increasingly thinned along a

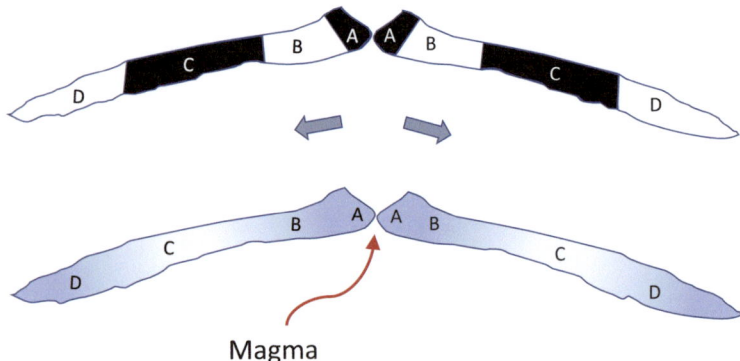

Fig. 2.13 The youngest areas of the oceanic crust are located at an MOR, where new magma intrudes and cools (A). Laterally, the crust becomes increasingly older (B to D). Fig. 2.14 explains why different configurations/ages of the Earth's magnetic field are stored at different locations of the oceanic crust. The black and white coloring schematizes the different polarities of the Earth's magnetic field recorded in the crust

---

[44]The average duration of times of the same polarity is about 450,000 years. In reality, there are much shorter and much longer sections. In any case, over a period of about 200 million years since the Jurassic, we will find several hundred polarity changes.

2.2 The Earth's Magnetic Field Over Time

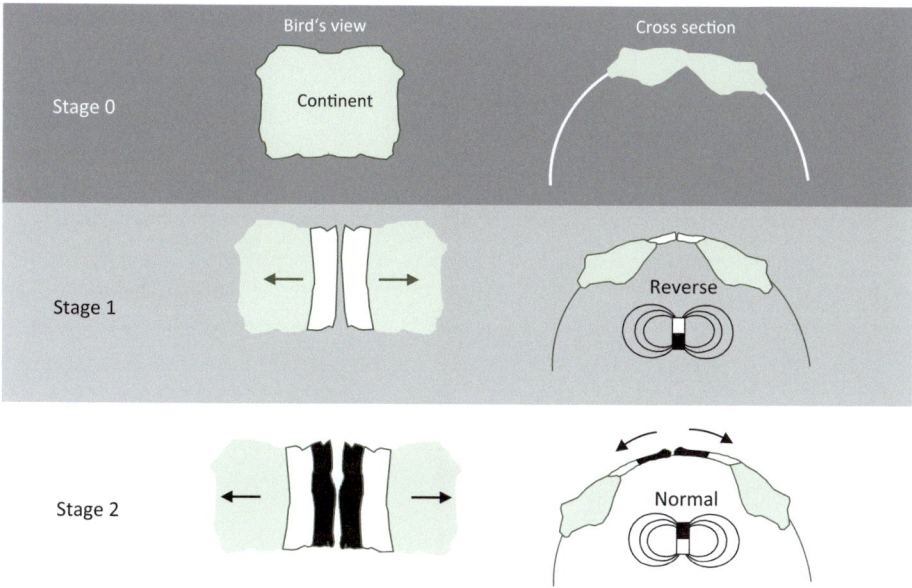

Fig. 2.14 Schematic stages of the formation and magnetization of oceanic crust. Stage 0: A continent from a bird's eye view before it breaks apart. Stage 1: New crust is formed from cooling basaltic lava at a mid-ocean ridge. Properties of the prevailing EMF are stored—here the magnetic north and south poles are reversed compared to today. Stage 2: More new crust has been formed by cooling, the continents have moved further apart. Since the EMF has flipped—i.e., north became south pole—this new configuration of the EMF is accordingly stored in the rocks of the oceanic crust

line due to tensile stress. It's as if dough is being pulled apart, only that the crust is less elastic and the thinning therefore mainly occurs through breaking. In geological terms, these are kilometers-long, deep-reaching faults in the earth's crust. Eventually, magma from the earth's mantle penetrates along the main fault line (the later MOR). Both sides of the breaking continent continue to move further apart, and the space in between is filled with rising and subsequently cooling lava. Oceanic crust is formed, which is topographically lower than the continent, flooded by seawater and transformed into an ocean over millions of years. This also holds an explanation as to why, in later plate collisions, older oceanic crust is usually subducted and thus no longer available for studies today[45] (Fig. 2.15).

---

[45] As a rule, the thickness decreases away from the MOR, as the crustal material cools more and more with increasing age and thus becomes denser, and consequently less thick (Fig. 2.15). Compare a loaf of bread fresh from the oven and a week later. Excluding local variations (such as so-called hotspots), the resulting surface of the oceanic crust from the MOR towards the continent resembles the shape of a suspension bridge from the central support pillar away towards the shore.

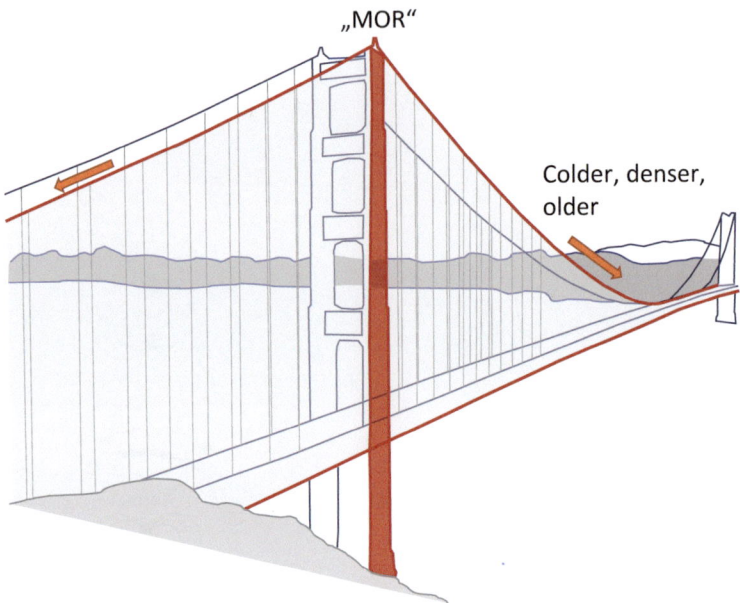

Fig. 2.15 Oceanic crust is youngest, hottest, and thickest in the area of the MOR. That's where the lava penetrates. In analogy with a suspension bridge, older and correspondingly colder, denser crust becomes increasingly thinner. Thinner and heavier oceanic crust is rather pushed downwards in a plate collision, subducted, which contributes to the disappearance of the oldest records of Earth's magnetic field in the oceanic crust.

Let's take a closer look. The earth's crust in the area of the oceans is about 6–10 km thick and is again made up of different layers. These different layers result from the chemical composition of a rock melt being continuously changed as it intrudes from the earth's mantle.[46] Both mentioned layers of the oceanic crust reach thicknesses in the 100-meter to kilometer range—certainly a topic for another book[47]. We just need to keep in mind that both the different composition and the cooling duration of a magmatic rock are important for the timing and durability of magnetization. And this has significance for the information on the earth's magnetic field in the earth's past, which we are looking for.

---

[46] In geology, magmatic rocks are named differently depending on the time of precipitation of their constituting minerals from the melt. For example, a *gabbro* crystallizes at an earlier time, more precisely at a different depth in the earth's crust and thus under different pressure and temperature conditions, than a *basalt*. Accordingly, both rocks are characterized by different properties, such as the type and amount of minerals contained and thus their appearance, their hardness, their mass and more. It is also easy to understand that magma, which comes into contact with cold seawater (due to the external shape called pillow lava), cools faster and differently than a rock melt cooling rather slowly in the upper areas of the earth's crust.

[47] For example, Frisch, Meschede and Blakey, 2010, *Plate Tectonics*, Springer.

In relation to the earth's radius of about 6370 km, the earth's crust is, as already mentioned, only as thin as an eggshell (Fig. 2.1). Since the earth is not an ideal sphere, but a slightly pear-shaped geoid, the distance from the sea surface to the center of the earth's interior varies by up to 20 km. The visualization of the earth's crust including the continents and ocean floors as a thin, movable and repeatedly breaking shell of our earth may make it easier to understand the dynamics of the earth's crust in the plate tectonic sense.

Other books deal scientifically in detail with the details of geodynamics. The topic ranges, for example, from the driving forces of plate tectonics to the geologically and geophysically complex processes in the area of the collision zones of the earth's crust (subduction zones). An original assumption from the first half of the 20th century, that the continents drifted like ice floes through the oceans[48], is far too simplistic according to today's knowledge. However, the underlying realization from this time, *that* continents have moved on the earth's surface in the earth's history, cannot be rated highly enough in its significance for the geosciences.

The explanation of how these large-scale movements within the earth's crust take place in detail, what drives the movements, how oceanic crust is newly formed at MOR and plays a role in geodynamics at least equal to that of the continents[49], has since claimed many years of international geoscientific research.

Certainly, there were other indications of the movement of the Earth's crust and, in particular, the continents. For example, continents that fit remarkably well on both sides of the oceans (Africa and South America!)[50], matching sequences of rocks or lines of flora and fauna on continents that are now separated, traces of movement as evidence of past glaciations, whose patterns only make sense if continents are pushed back into a previous position, etc.

However, alternative explanations for such findings were repeatedly devised. Even if the outlines of Africa and South America fit well together when both continents are moved back towards each other by reversing the development of the ocean—Madagascar also fits remarkably well into the Black Sea. However, in this latter case, it would not be possible to explain how such plate tectonic movement could have occurred without air travel.

In particular, no comprehensive theory on the dynamics of the Earth's crust as a whole could be presented for a long time. The magnetic stripe pattern on both sides of the MOR—symmetrical and independent of the structure of the oceanic crust itself—was able to make a significant contribution in the 1960s: A decreasing age of the crust from the MOR to the continent, measured geophysically, was difficult to refute in its

---

[48] Continental Drift, named by Alfred Wegener (1880–1930).

[49] For example, Stüwe, Kurt, 2010, *Geodynamics of the Lithosphere,* Springer.

[50] For example, almost 60 years ago, using numerical methods, the so-called Bullard's Fit: Bullard, E., Everett, J. E. and Smith, A. G., 1965. The fit of the continents around the Atlantic, *Philosophical Transactions of the Royal Society of London,* Series A. Vol. 258, No. 1088, pp 41–45.

significance. Why wasn't the knowledge of the magnetic stripe pattern used earlier? A look at the development of the GPTS makes it clear that the magnetic timescale was not established in any way in the 1950s! A few years earlier, the existence of magnetic pole reversals was not even scientifically proven, let alone measured in rocks and assigned to a timescale. The scale known today is still constantly being refined and expanded, especially in the difficult to investigate beginnings of Earth's history in the Cambrian and Precambrian. However, the existence of the pole reversals is now proven, as is the possibility of recording these reversals in rocks.

But let's take a closer look again before the idealized thought of a wonderfully cleanly drawn barcode in the ocean floor unreflectively takes hold in our minds. On a sphere with a rigid surface, here the Earth with the Earth's crust, a linear element, here the MOR, hundreds of kilometers long, will always have to be offset by approximately right-angled transverse fractures (so-called transform faults in the geological sense) to accommodate the curvature of the (rigid) spherical surface (Fig. 2.16). In addition, magma will not intrude everywhere at exactly the same rate over a distance of hundreds of kilometers, and there will always be secondary lines of flowing lava, so that the supposed straight line, when viewed more closely, has an irregular shape in individual sections separated and offset by faults.

For the MOR, a hint at the actual dimensions of such an underwater zone should clarify the reality beyond our simplifications. What we have represented as a simple line in Fig. 2.16 is comparable in reality to an underwater mountain range, sometimes with a valley zone with lateral "heights" often rising hundreds of meters. Not along all MORs is the ocean floor created at the same average speed, and the described morphology of the MOR is strongly associated with this spreading speed. Speeds are on the order of 5 cm (slow) to 15 cm (fast) per year, equally on both sides of the MOR. If we compare, for example, the magnetic stripe patterns from areas of the Atlantic and the Indian Ocean, we expect the same stripe pattern for oceanic crust, which was created at the same times in Earth's history. Overall, the beginning opening of the Indian Ocean is the younger geological event; accordingly, the oldest magnetic stripe patterns do not go back quite as far in Earth's past as in the Atlantic. From the youngest Jurassic period to the present, the magnetic pattern can be compared, although the stripes are relatively elongated or compressed according to the different spreading speeds of the Atlantic and the Indian Ocean[51].

Overall, the magnetic stripe pattern thus shows a significantly more complex stripe pattern than is suggested by the term "barcode". The magnified view from Fig. 2.16 of a part of the MOR in the area of a transform fault illustrates the complexity and

---

[51] For example, 100,000 years of normal polarity at a spreading speed of 10 cm/year produce positively magnetized lava of 5 km width on both sides of the MOR. At 5 cm/year, *the same period* in the image as magnetized Earth's crust is shortened to two stripes of only 2.5 km width each.

2.2 The Earth's Magnetic Field Over Time

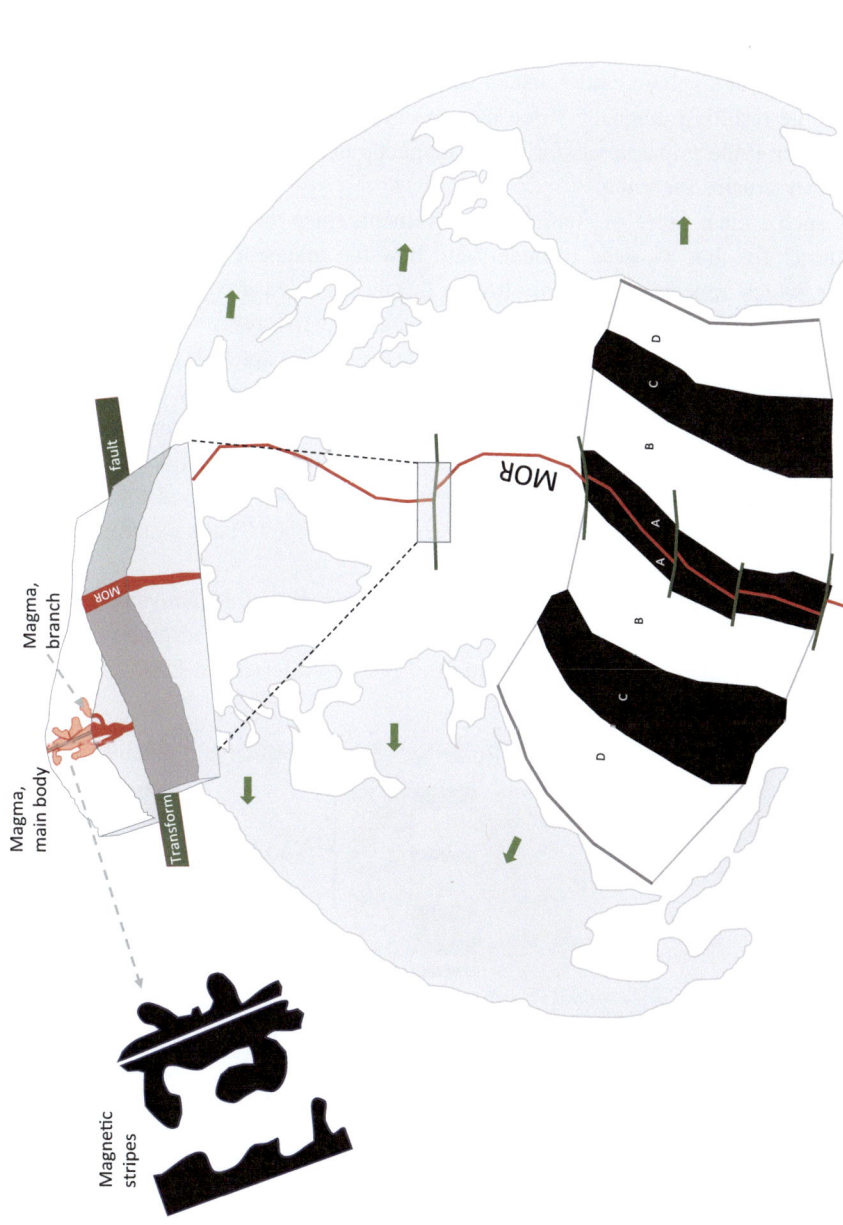

Fig. 2.16 Schematic representation of a MOR. From a bird's eye view a MOR is a linear, elongated feature, which is laterally offset by transverse transform faults. These faults also affect the distribution of any previously stored magnetic stripe pattern. A three-dimensional zoom into such a junction of MOR and transform fault illustrates that magma does not always intrude "orderly", but quite often in side branches away from the main zone of a MOR. On closer inspection, the magnetic stripes of the oceanic crust have "irregular" edges (top left; see also Fig. 2.12, 2.13 and 2.14)

the reasons for the deviation from an ideal stripe pattern as a result of local geological variations.

A structure the size of an ocean does not open at every point of the hundreds of kilometers long fault zone (of the MOR) at the same rate, along an ideal loxodrome on the surface of the earth's sphere, and constant over the millions of years. Accordingly, we expect that the resulting magnetic stripe pattern deviates from the ideal barcode across the ocean. Even if the principle of storing the magnetic field at the time of the cooling of the rising lava remains the same.

So not such a clear proof of continental movement, since there is no clear magnetic stripe pattern? Yes, but we need to understand how the magnetic stripe pattern of the ocean floor, which appears simple only from a distance, is measured. Should individual rock samples be drilled and, if so, how often? Is there another way?

If all the black and white stripes of the barcode are to be captured, then at least one sample—a measuring point—from the respective stripe is necessary to capture the polarity of a stripe (Fig. 2.17). When sampling along a traverse, an equidistant sample distance is usually chosen to avoid further complicating the investigation. The sample distance should be chosen so that the narrowest white or black stripe of the pattern is wider than this distance. But how wide is the narrowest stripe? That's exactly what we want to find out first.

To this end, consider the following: From a spreading rate of 10 cm/year, a normal or reverse polarity of 100,000 years duration results in a 5 km wide stripe on each side of the MOR. Remember: The average duration of polarity intervals in Earth's history is

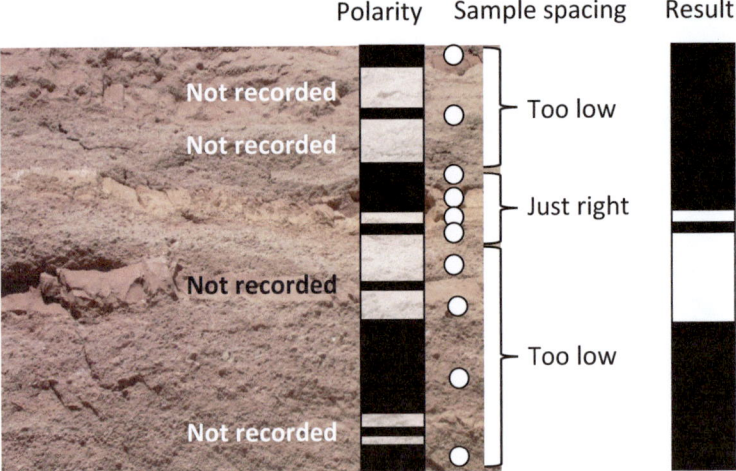

Fig. 2.17 A hypothetical polarity pattern is projected onto a sequence of rock layers (left barcode). Since the polarity changes are independent of the layer boundaries, other geological considerations must be made to realize a sample distance (circles) as adequate as possible relative to the frequency of the black-white changes. A too small distance carries the risk of not capturing all polarity changes. This would obviously lead to a faulty and thus misleading barcode (right barcode) (see also Fig. 4.12)

450,000 years, the longest chrons lasted several (tens of) millions of years, the shortest only a few thousand years. The mean value can therefore only be a first indication. In our example, with a sample distance of 5 km, all chrons shorter than 100,000 years would often not be captured. Or if by chance they are, then the resulting black or white stripe would be displayed wider than the actual duration of the chron. The reason for this is that the boundary between two samples of different polarity is drawn halfway between these two samples. In other words: halfway between the last information about normal polarity and the first information about reverse polarity. In our case, the boundary would be 2.5 km from the respective sample. However, it would be extremely laborious to drill a continuous profile of individual samples from the ocean floor, measure these samples and then analyze them. The water depth alone is over 4000 m over large parts of the profile.

While this is fundamentally no problem for a research vessel like the *JOIDES Resolution* of the IODP (International Ocean Discovery Program; further explanations in Chap. 4), in our example about 600 laborious drilling stops would be necessary if the traverse was planned over the 3000 km from the MOR of the Atlantic to the coast of North America. That would be far more drillings than this research vessel contributes in dozens of missions over a period of five to ten years with correspondingly many different interdisciplinary geoscientific research topics for understanding our Earth. So not a good cost-benefit ratio.

Fortunately, there is another way to magnetically measure the ocean floor, namely by attaching a magnetometer to a ship and dragging it across the ocean (Fig. 2.18 and 2.19). Of course, this terse statement does not do justice to the actual physically and technically complex execution of such an experiment—in essence, however, it is actually measured whether the magnetic signal of the oceanic crust adds to the magnetic signal at the

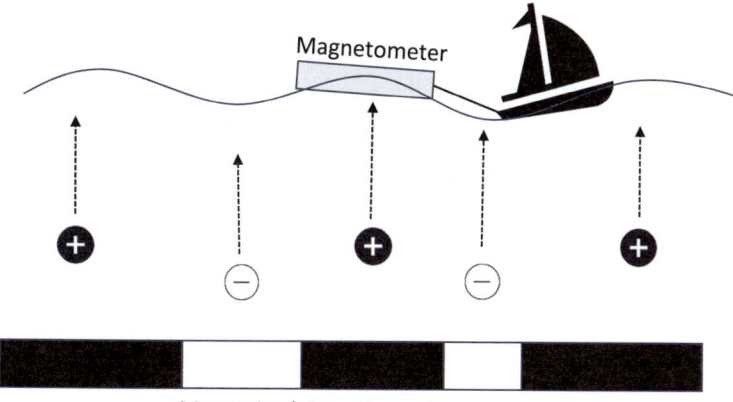

Fig. 2.18 Schematic representation for measuring the magnetization of the oceanic crust. A magnetometer at the water surface registers increased or decreased values, depending on whether the crust is magnetized *in* or *against* the direction of the current Earth's magnetic field. In this way, the magnetic stripe pattern of the ocean floor is mapped.

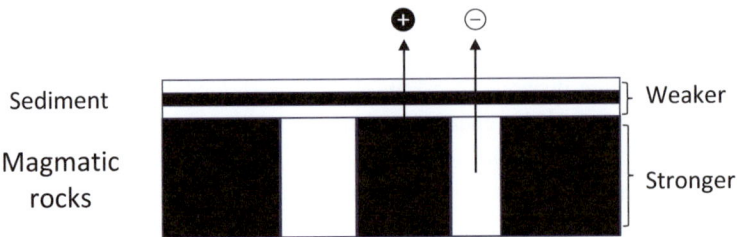

Fig. 2.19 In some geological scenarios, sediments may cover the magmatic rocks formed at the MOR. Such sediments also record the prevailing direction of the Earth's magnetic field when they are deposited. The magnetization of these stripes is thus "perpendicular" to the magnetization of the magmatic crust. Individual layers may thus cancel out their contributions to our ship magnetometer, but overall they are much weaker than the underlying magnetized magmatic Earth's crust.

surface (black, normal stripe) or subtracts (white, reverse polarity). The latter depends on how the rock is magnetized at depth.

Overall, the magnetization of both the rock and the Earth's surface is extremely weak by human standards; however, correspondingly sensitive magnetometers can easily pick up the signals.

Drilling individual rock samples, as initially mentioned, is a common practice in paleomagnetism. But only if either geological profiles on a rock wall or larger drill cores, such as those pulled by a research ship, are sampled. So if you've ever wondered about a number of circular holes about 2 cm in diameter in a natural rock wall, you've probably been looking at the traces of a paleomagnetic sampling.

Especially in the natural sciences, the consideration of adequate dimensions is crucial for understanding a matter. In this respect, the ultimate averaging of the results through the methodology with the magnetometer in tow is rather advantageous for analyzing the deviation of the magnetization of the ocean floor from the ideal stripe pattern. To average out the effect of lateral offshoots of rising and cooling lava, several parallel traverses of the ocean floor would otherwise have to be sampled. In other words: To understand the extent of a piece of forest, it is not necessarily necessary to measure the diameter of the individual tree trunks.

Searching for Stripes on Land

So far, the recording of the magnetic stripe pattern of the first 200 million years of Earth's history—supported by the analysis of actually drilled rock, the barcode of the Earth's magnetic field is well documented. But what about the older sections of Earth's history?

Here we have to rely on a mosaic of many individual continental outcrops. This work is much more laborious and possibly tedious, as such studies are usually associated with the drilling of the mentioned about 2 cm large rock samples. Many well-placed and well-documented rock samples to meet the requirements for accuracy and statistical quality criteria of a paleomagnetic investigation. Especially in difficult terrain, a team is

2.2 The Earth's Magnetic Field Over Time

often needed to transport a suitable drill including non-magnetic hollow drill bits, cooling water and a generator to the site of the sample collection and later, after the work is done, to transport the rock samples back.

The actual difficulty with continental profiles, however, lies elsewhere, because ideally the following criteria should be met for a successful paleomagnetic investigation:

- Continuous sequence (sounds uncomplicated, but due to the expected discontinuities[52] easier said than found)
- No tectonic stress (also not trivial; in the simplest case this includes a tilting or deformation of the layers up to a metamorphic change of the original rock caused by high pressures and temperatures)
- Stable magnetization (Chap. 3)
- For sediments, an appropriate accumulation rate to achieve adequate temporal resolution (Fig. 2.17)
- Accessibility (see above regarding sampling)
- Several such outcrops worldwide for securing and confirming found reversals of the Earth's magnetic field
- Dateable with alternative methods outside of magnetic investigations

With *continuous sequence*, a rock is labeled that has formed over a period of—ideally—several million years in a well-understandable process (that's why we find the oceanic crust so suitable). Whether sedimentary or magmatic, it is essential that the Earth's magnetic field was recorded continuously over the formation period of the rock. For cooling lava at a MOR, we have already discussed these criteria. But now we are looking for rocks that are older than the oldest oceanic crust. Sediments **on** the oceanic crust are understandably not older than the underlying oceanic crust, so we are left to examine the continental area. And on land, we will mainly have to look for sedimentary rocks to sample a longer geological sequence.

The reasons for this are: There are many magmatic rocks in the continental environment, such as volcanic lavas or ejected ashes. However, volcanoes that are *continuously* active over long periods of time are rare—fortunately, one should add in this case, at least with regard to the flora and fauna of the time. We would therefore expect records of the Earth's magnetic field through magmatic rocks to be rather temporary and fragmented. In the best case, one will be able to examine magmatic rocks that solidified at a deeper location in the Earth's crust and are now exposed through uplift and erosion.

We also place less emphasis on metamorphic rocks in our search for suitable magnetic recorders. After all, metamorphosis in the geological sense involves the action of increased pressures and temperatures on rocks—processes that are very likely to destroy

---

[52] There are some geological examples to follow in Chap. 4 and 5 to explain the notorious "irregularity" and "interruption" in the formation of rock layers.

a previously acquired original magnetization of a rock (Chap. 3). Metamorphites can also carry a record of the Earth's magnetic field. This is acquired after the increased pressures and temperatures have subsided, usually after the end of a tectonic, for example, mountain-building phase. The recorded magnetic field thus reflects the age of the "mountain formation", not the age of the initial formation of the rock. And is certainly independent of the original rock boundaries.

Metamorphosis in general—anyone who delves into this field of research will probably rightly turn up their nose at the seemingly casual designation in the face of this geologically, mineralogically, and physically complex area—usually fundamentally changes an existing rock. An original layering is often superimposed in the course of metamorphic processes and replaced by one or more *foliation* surfaces, often running at an angle to the former layering. However, layer surfaces are an important marker in rocks to further narrow down the period of formation of a rock. We will make various considerations to understand the rate at which sediments or magmatic rocks form in different geological scenarios. In short, rock boundaries will serve us as an important temporal framework for dating rocks.

Foliation surfaces, on the other hand—here used for simplicity as a collective term for a variety of possible types of metamorphic separation surfaces in rocks—can only fulfill such a framing function in the dating of rocks to a very limited extent. Let's imagine foliation surfaces as mineral grains rotated into a uniform orientation by rock pressure in a (former) sediment. While the original layers were formed by *gradually* deposited mineral grains, the foliation surfaces were formed almost *simultaneously* in a larger volume of rock, as our assumed mountain-building process obviously affected a large volume of already existing rock. The crucial point now is: Corresponding to the foliation surfaces, the recording of the Earth's magnetic field—with the decrease of pressure and temperature—will also occur almost *simultaneously*.

In a figurative sense and with regard to the polarity reversals of the Earth's magnetic field, an original sequence of normal and reversed magnetization in sedimentary rock is replaced by a uniformly "white" or "black" sequence (our graphic translation of normally and reversely polarized sections of Earth's history) in the now metamorphic rock. A less striking view may also yield a refined temporality in the recording of the Earth's magnetic field for metamorphic rocks under certain circumstances. However, this will not change the fundamental limitations of working with metamorphic rock. An event (end of metamorphosis) is (magnetically) recorded, not a geologically long, continuous process of rock formation.

Sediments remain for the seamless recording of the Earth's magnetic field. Possibly also a combination of such sedimentary rocks and intercalated volcanic layers. In the context of Fig. 4.3 to 4.8 we will make some considerations about possible sedimentary processes and their (geological) rapidity. At this point in our considerations, it is important to emphasize that we want to create the geomagnetic standard scale first—and do not already have it available for dating. In other words and another analogy: We are exploring new territory and not navigating with an already created map.

2.2 The Earth's Magnetic Field Over Time

As so often in this book, we cannot discuss all possible geological scenarios. However, we can think of examples to understand geological scenarios of a long-lasting recording of the Earth's magnetic field.

Let's assume that the recording is done by sediments, i.e., in the sense of a detrital remanent magnetization to be discussed in Chap. 3. Ideally, sediment material will then be continuously supplied over a geologically long period of time, which records the Earth's magnetic field during the course of deposition over this period. The amount of material supplied—the accumulation rate—should be just high enough to capture "relevant" changes in the Earth's magnetic field. If a new, thin sediment layer is formed only every 100,000 years, for example, a short polarity reversal of 20,000 years would not have been recorded. Or rather, the magnetic reversal would not have been resolved in a magnetic measurement, as the rock sample (in terms of time) would be larger (longer) than the recorded polarity change.

Two extreme examples of sedimentary accumulation rates are lake sediments (very low) versus delta deposits with a large, easily erodible catchment area of the supplying river (very high rate). For example, in the vicinity of a geologically young, still "rising" mountain range (Bangladesh/Himalaya), extreme amounts of sediment material are transported into the plain, while in a low-flow lake in the high mountains, only what was blown onto the water surface by air (aeolian) sinks to the lake bottom (Fig. 2.20).

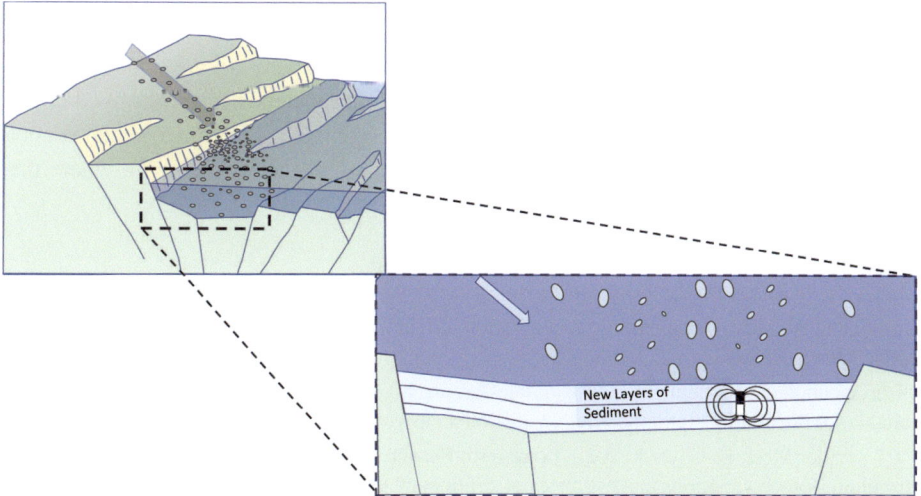

Fig. 2.20 Example of a geological situation in which clastic sediment is delivered from the surrounding highlands and deposited in a basin. The basin proposed here for the illustration of continuous sedimentation is defined by deep-seated faults in the Earth's crust. The faults could, for example, compensate for a stretching of the crust at this location. With the deposition of new material, the prevailing Earth's magnetic field is recorded in the course of a detrital remanent magnetization. Consideration for a possibly long-lasting continuous sedimentation: At some point, the basin would be filled up, and sedimentation would stop. Unless the stretching of the Earth's crust and thus the subsidence of the basin is not completed and active at a rate balanced with the delivery of new sediment material. Such geological situations do exist—and thus also corresponding records of the Earth's magnetic field over a long period of time

Fig. 2.21 Chalk cliffs of Dover. Over many millions of years, the calcareous components of algae and other organisms have accumulated on the former seabed during the Cretaceous period—layer by layer and horizontally deposited. Lower sea levels have exposed the seabed, and erosion, especially since the last ice age, has shaped the current cliffs along the English Channel

With these considerations in mind, carbonate rock[53], in a tectonically quiet setting, is an ideal candidate for continuous sedimentation and chemoremanent magnetization (Chap. 3). An image to visually underline these considerations might be the horizontal layers of marl on the coast of Sicily or the chalk cliffs of Dover[54] (Fig. 2.21).

In addition, we must note at this point that rocks exposed today in the continental realm do not necessarily have to have been deposited or erupted in a continental environment. Often, the sea level plays a crucial role when marine deposits were formed on formerly flooded continental crust and are now exposed on land. Imagine a sea level about 200 m higher than today (actually comparable to the Cretaceous period[55]), where then land below 200 m above sea level is flooded.[56]

---

[53] Sediment formed by evaporation, precipitation, or simply the deposition of calcareous algae and corals in tropical seas.

[54] The mentioned sediments in both examples are each younger than the oldest oceanic crust, but suitable for illustrating similar, earlier depositional environments.

[55] Cf. van der Meer, D. G. et al., 2022. Long-term Phanerozoic global mean sea level: Insights from strontium isotope variations and estimates of continental glaciation, *Gondwana Research,* Vol. 111, pp. 103–121.

[56] From a perspective of ongoing discussions, it should be noted: Scientific studies and forecasts on the influences and effects of human-induced climate changes take into account natural changes in the environment in geological history. The influence of humans on system Earth then manifests itself as an overlay and often significant alteration of already known, natural fluctuations of system Earth. Precisely *because* we have a now quite detailed picture of hundreds of millions of years of geological Earth history from decades of geoscientific research, we can assess what new and

2.2 The Earth's Magnetic Field Over Time

As further explained in Chap. 4, tectonics can affect the recording of the Earth's magnetic field in different ways. In the scenario described here, sedimentation and tectonic movements occur simultaneously. Syntectonic. The magnetic signal can usually be recorded without problems in such geological scenarios, especially when we imagine that such tectonic movements, like sedimentation, proceed very slowly. Imperceptibly—should an observation in real time be attempted. None of this stands in the way of the acquisition of a detrital or chemoremanent magnetization.

No tectonic stress should therefore be more appropriately termed: *no subsequent* tectonic stress. If once deposited sediments (or any other rocks) are subsequently displaced, offset against each other[57] or even folded, this also means more additional paleomagnetic work rather than a destruction of the magnetic signal. Basically, in such cases, in addition to the magnetic analysis, a geometric puzzle needs to be solved. The correct temporal sequence of the sediments must be achieved by correctly "ordering" the packages. An example of this is shown in Fig. 2.22.

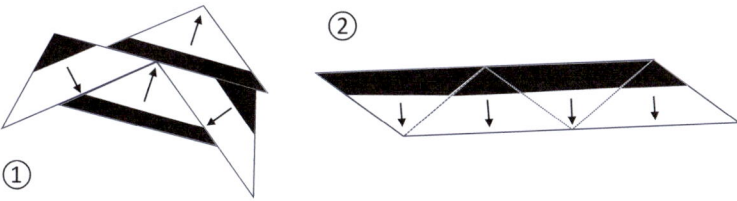

Fig. 2.22 Highly schematized, ① shows subsequently offset sediment packages. The arrows symbolize a recorded magnetization direction before tectonic stress. A purely geological approach would describe several faults/fracture surfaces in the rock. Schematically, these are the outlines of the triangles here. Two of the triangles would be described as "upside down". We assume that the white layer was deposited first and then the black layer. Typical sediment structures can show something like this. Magnetic measurements would indicate different directions of the recorded Earth's magnetic field according to the arrows. If the tectonic movements ② along the faults are reversed (for simplicity's sake on paper, not on the rock itself), a uniform sediment sequence, including a uniform magnetization, results. This sequence represents the layers as they were originally deposited and magnetized before the tectonic disturbance.

---

often climatically relevant factors have been added in the last 100 years. And these factors can be clearly traced back to human activities. Geological factors such as the plate tectonic movements and a resulting reconfiguration of the continents mentioned as examples may cause more dramatic environmental changes—200 m water level change (!) in the Cretaceous—but this happens over periods of many millions of years and not in the course of a human life or even several generations. Even a 1 m higher sea level, in addition to natural fluctuations, has devastating effects in the coastal regions of our densely populated Earth. And a changed sea level is known to be just one of many possible effects of an unnaturally changed climate on Earth.

[57] An example is Fig. 4.5 in the discussion of a geological scenario.

As a visualization aid, one can imagine the direction of the Earth's magnetic field recorded in the rocks as an arrow (more on this in Chap. 4). Zoomed out, this arrow points parallel to the field lines of the Earth's magnetic field at this location on Earth at the time of rock formation (Fig. 2.22 and 4.13). If one were to take a chunk of rock in hand and turn the rock out of its original position (tectonics in the geological sense), the (imaginary magnetic) inscribed arrow would also be turned out of its position in the same way.

While tectonics thus causes additional work for paleomagnetic studies to establish the polarity time scale, it should be remembered that the necessary geological markers for clear reconstruction are not always present in the rock. In such cases, a rock sequence is not ideal for this type of pure paleomagnetic research, as the geometric reconstruction of the puzzle pieces is not ideal due to missing reference points for the geological sequence.

The recorded magnetic signal (no tectonics!) really becomes problematic in areas of tectonic stress on the rock, where the rock has been internally altered mineralogically and structurally. This can happen due to the influence of increased pressures and temperatures during the deformation of the Earth's crust. As an example, one might consider that in the apex area of a rock fold, the stress on the rock is stronger than on the flanks of a fold (Fig. 2.23). While magnetic samples from the flank area could be used, a sample from the apex area may have been altered too much. In such cases, the magnetic recording was also destroyed. In the extreme case, tectonics lead to a metamorphic change in the rock, as already discussed in the previous paragraphs. Such layers are discarded as suitable rock for creating the polarity timescale.

These rejection criteria are almost self-explanatory, because for paleomagnetic studies, rock samples must be taken from (in geological terms) outcropping layers. Cleanly and precisely drilled sample cylinders of about 2 cm in diameter are one of the common

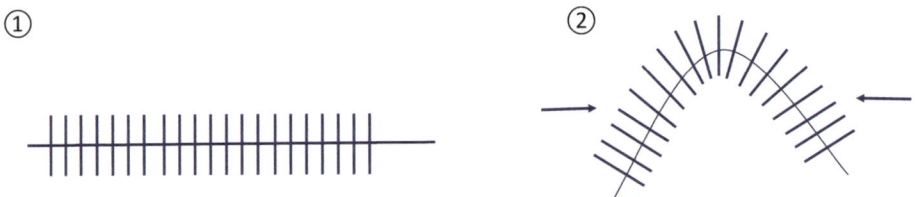

Fig. 2.23 In the apex area of a (geological) fold, the elongation and stretching of the material (a rock layer) are most pronounced. The schematic drawing compares an undeformed layer ① with a fold ② that has arisen from this layer due to lateral pressure. The distances of the short parallel lines in ① change most noticeably at the apex of the fold (further or closer respectively). For simplicity's sake, we assume a deformation of the layer solely by stretching and compressing, but without breaking. Depending on the nature and type of rock layers, other deformation mechanisms are possible and likely.

methods.[58] To avoid magnetically contaminating the samples, special hollow drill bits made of non-magnetic material are used. Drilling is done with water for cooling. Equipment including the water canisters must be transported to the sampling site and used safely.

Ultimately, the determined magnetic direction should also be described by as precise angular specifications as possible (declination and inclination; Fig. 3.11). Such angular specifications are made in relation to a reference system—usually the geographical grid of longitude and latitude—and therefore a paleomagnetic rock sample must be marked before it is taken from the outcrop. Every degree of inaccuracy will be transferred to the final paleomagnetic result. Such a marking could be an arrow inscribed on one side of the sample, which is measured with a compass. In other words: It is ensured that the original three-dimensional position of the sample in the rocks can be precisely reconstructed, even if the sample often lies many hundreds of kilometers away on a laboratory table.

At this point, we should realise that paleomagnetic sampling in difficult terrain must also ensure oriented sampling. Imagine a supposedly ideal profile on a steep wall in the high mountains below an overhang. Sampling is not impossible, but meeting the criteria mentioned so far certainly requires increased logistical effort.

Why should several outcrops be examined? Here we refer less to quantitative acceptance through mass compatibility in everyday language than to pure quality, which therefore finds confirmation. In the scientific sense, i.e., for a reversal event of the Earth's magnetic field in Earth's history to find its rightful place in the paleomagnetic polarity timescale, simultaneous polarity changes at other locations on Earth with the same quality features must be proven. After all, in our considerations of the Earth's magnetic field, we assume a global structure. We have already discussed that the Earth's magnetic field will have different angles of inclination at different points on Earth. No matter where we are on Earth, a polarity change will always turn the local magnetic value upside down, so to speak. And this should have been recorded at various locations on Earth. Ideally, because as said, there are periods in Earth's history for which almost no—at least not paleomagnetically suitable—rocks are preserved. Sometimes quality (of a magnetic recording of the Earth's magnetic field) will also have to seek confirmation.

---

[58] Why 2 cm cylinders? In the following chapters, we will find more clues. All in all, it is sufficient for this book to say that magnetic measurements depend in different ways on the amount of material examined. If we imagine the stored magnetization directions of the Earth's magnetic field as a swarm of more or less parallel, small arrows in the rock, a too small sample may not contain enough small arrows to make a statistically relevant statement about the common direction of these arrows. 2 cm is a good compromise considering the resolution and practical size of the measuring devices, the average amount of "arrows" in rocks, or also a practical size for taking samples. Cylinders are also a fairly easy to produce sample shape (hollow drill bit) and also quite close to the ideal shape of a sphere for the samples. Spheres provide the same amount of material in every spatial direction during a measurement, but are harder to produce, label, or store.

Once established, the magnetic polarity timescale is primarily intended to be used for dating sequences of rocks of unknown geological age—in such cases, we ideally will not have to ask for another dating method. However, in order to initially establish the polarity timescale as a standard, it is essential that a period of time can also be verified by non-magnetic or, more generally, independent dating (Fig. 2.24). We have already mentioned that a variety of dating methods are used in the geosciences. These methods are based on different procedures, from the use of radioactive decay series to fossils to cyclical changes in the Earth's orbit. All these methods have in common that no methods cover the entire period of geological Earth history in the same way, i.e., with the same accuracy. For example, the half-life of the carbon isotope ^{14}C relevant for radiometric dating is 5730 years, while the same value for uranium (^{238}U) is 4.47 billion years. Without going into the details of radiometric dating,[59] it can be inferred from the comparison of the two half-lives that the mentioned isotopes can be used for geologically young and—even in geological terms—old periods of Earth's history respectively.

Almost even more important for the practical application of these methods is to consider that a rock, in order to be dated, must also contain the corresponding radionuclides.

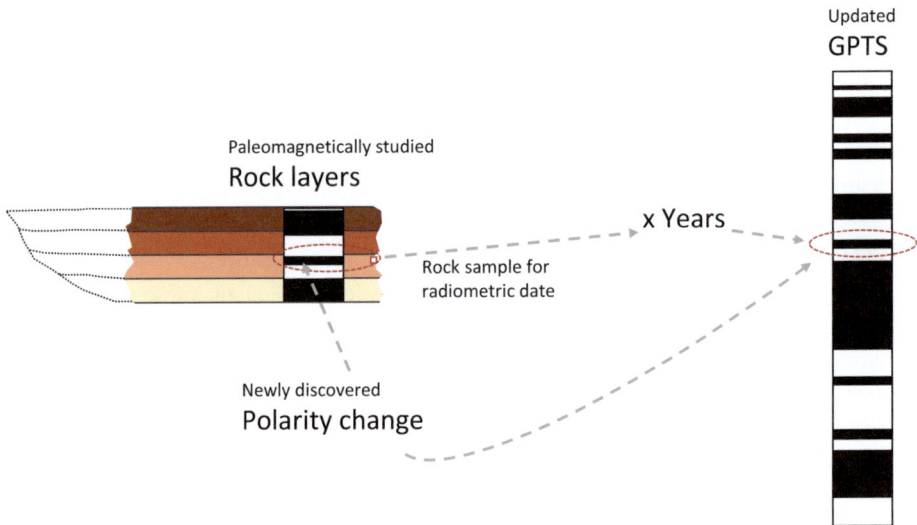

Fig. 2.24 In this example, we assume that paleomagnetic measurements of a rock sequence, taking into account all necessary geological quality criteria, show a previously unknown polarity reversal (here normal polarity, black). The determination of the absolute age of the beginning and end of this reversal is then part of the implementation into the standard GPTS. This age determination is usually achieved through suitable radiometric dating.

---

[59] For those who want to read up, see for example I. Wendt, 1986. Radiometric Methods in Geochronology, *Clausthaler Tektonische Hefte* 23, Springer.

2.2 The Earth's Magnetic Field Over Time

This means in practice, especially for rocks from the older Earth history, that they contain certain minerals which can be extracted for dating. For example, zircon or feldspars are two typical minerals used for some radiometric dating. These minerals are usually not contained in pure carbonate rocks (such a rock could, for example, consist mainly of shell fragments).

Furthermore, it is important that during the supposedly million-year history of the rock, no disturbance of the decay series of the radiometric isotopes (for example, due to tectonic processes) has occurred, which would falsify the result. Therefore, it may be necessary to select a different, non-magnetic dating method for a 300 million year old polarity reversal than for a 2 million year old one, and to adapt this dating method according to its suitability for the type of rock.

So let's summarize: The GPTS—this paleomagnetic standard barcode—has been continuously expanded and refined since the realization of the repeated reversal of the Earth's magnetic field in the 1920s. Several hundred polarity reversals have now been recorded for the time since the Cambrian, and each of these polarity reversals is correlated with the geological timescale through an absolute (often radiometrically determined) age. The irregularity of the polarity reversals throughout Earth's history has created a pattern of black and white periods, which does not show a comparable pattern at any point in Earth's history and thus delivers a unique signature comparable to a fingerprint. The examples mentioned above show that the GPTS is also a dynamically developing object. Even though the basic structure is now quite well known and can be considered secure, refinements of the absolute age of individual polarity reversals will always be made in certain areas. It is also likely that previously unknown polarity reversals will have to be found and integrated into the GPTS in certain magnetically disadvantageous periods of Earth's history. We have already discussed that magnetically disadvantageous periods have less to do with the quality of the Earth's magnetic field than with the lack of a suitable recorder. Geologically speaking, there have always been phases in Earth's history when—even on a global scale—no or only a few rocks were formed that could have recorded the Earth's magnetic field. Or recorded directions of the Earth's magnetic field have largely been destroyed in phases of preferential convergent plate tectonic reorganization. In other words: For example, sediments of the continental shelves have been folded and metamorphosed during mountain building phases.

One more note on the practical application of the GPTS: We have also already mentioned that the current, normally polarized section of Earth's history was named after Bernard Brunhes. In order to be able to navigate through the hundreds of older (black and white) polarities of Earth's history that have been discovered since then, each of the stripes must be named and assigned according to its position in the timescale of Earth's history. Even if it were possible to name a corresponding number of deserving paleomagnetic researchers as namesakes, such an approach would obviously be at least confusing. The naming procedure of choice therefore uses a—in principle—hierarchical, alphanumeric system.

Periods of Earth's history with predominantly the same polarity of the Earth's magnetic field lasting around 1 million years are referred to as a Chron. Shorter times on the order of 100,000 years are accordingly referred to as Subchron. Even shorter flips of the Earth's magnetic field of less than 30k years are referred to as excursions due to their short duration.

All polarity flips since the Jurassic period, which we can verify using the basalts of the oceanic crust as described, are accordingly divided into two large sections: the C- and from the Cretaceous onwards the M-sequences. For example, we refer to the Brunhes-Chron as C1n—from today's perspective the first normal polarity of the Earth's magnetic field -, the subsequent Chron as C1r[60]—the Matuyama-Chron is thus the first reverse polarity from today's perspective. The number increases with the increasing age of the Chron. Just before the Cretaceous period, we have reached C34, from then on the M-Chrons begin. If a subchron is discovered later in the Brunhes-Chron, this is referred to as C1n1r (first inverse polarity within the—predominantly—normally polarized Brunhes-Chron). If we were to further subdivide C1n1r, the discovery of another normal section (or, better said, the realization that the inverse interval 1n1r actually consists of two separate, shorter intervals), this would be referred to as C1n1r1n. That's the idea. However, even in the case of the GPTS, the reality of developments sometimes forces compromises, so that some of the numbers are missing or some Chrons have received the additions A, B or C[61].

All known polarity changes older than the Jurassic have also been named, but still need to be integrated into a more stringent nomenclature in the course of further research into these older states of the Earth's magnetic field. Overall, it is important for us as readers at this point to understand that each stripe of the GPTS barcode can be clearly identified and named.

So what does a typical dating using the GPTS look like? How does magnetostratigraphy finally work?

In Chap. 4 we consider a fictional geological scenario, which over a period of several million years leads to successive deposition, partial erosion and tectonic displacement of the rock layers in a limited area. Following this geological excursion, we will then drill several samples in this fictional depositional environment for paleomagnetic investigations. Of course, the sampling is also only theoretical, but one can see how the Earth's magnetic field helps us to date this fictional sequence.

However, first we need to take a closer look at magnetism and magnetization of rocks in Chap. 3.

---

[60] In German usage, "invers" becomes "reverse" in Anglophone usage, hence the r.

[61] Just as in a street sometimes house number B is inserted, for example when one of the plots has an additional new house built. To collect all house numbers and renumber them would usually involve much more effort, including a new postal address for each resident and corresponding re-registrations.

Magnetism

3

First, we want to understand the principle of how rocks can store a magnetization of the Earth's magnetic field at all. The principle.[1] But even for this, we need to look more closely. In the direct sense, down to the subatomic level, i.e., to the components of the building blocks of minerals. In a figurative sense, for example, to our understanding of magnets.

So what is fundamentally important for the magnetization as well as the consideration and application of the magnetization of rocks? Let's start with a more or less unsorted list of assertions. The explanations then follow accordingly sorted:

- All geological formations, i.e., all naturally formed rocks of the Earth, carry at least one record of a past state of the Earth's magnetic field.
- Not the entire rock, but only a few selected minerals record the Earth's magnetic field. Such special "magnetic" minerals have certain properties, at least at the level of their crystal lattice, in order to be useful for recording the Earth's magnetic field.
- Such minerals with magnetic memory, from a physical point of view, fall into the group of so-called ferromagnetics, but even among these, they still form a heterogeneous community (Sect. 3.1).
- The majority of minerals existing in and on the Earth's crust, on the other hand, belong to the groups of diamagnetics and paramagnetics. Such minerals do react to a

---

[1] Those interested in a purely scientific summary are referred to the relevant technical literature. An introduction to the physical theory and practice is provided, for example, by the book by Dunlop, D.J. & Özdemir, Ö., 1997. *Rock Magnetism, Fundamentals and Frontiers*, new edition 2010, Cambridge University Press, 573 pages. There you will also find citations to countless, fundamental individual publications on this topic.

© The Author(s), under exclusive license to Springer-Verlag GmbH, DE, part of Springer Nature 2025
M. Urbat, *Magnetism of Rocks*, https://doi.org/10.1007/978-3-662-70428-8_3

magnetic field, but they "forget" their reaction as soon as the field no longer acts. Out of sight, out of mind. These minerals are still magnetically interesting.
- With minerals with memory—remanent magnetic minerals in the scientific sense—not only the *type* of mineral, but also their respective grain size is a decisive factor for the permanent recording of the Earth's magnetic field. The rock particles should be neither too large nor too small (Sect. 3.3).
- The recording of the Earth's magnetic field takes place at the time of the formation of the rock and remains preserved in the rock over geological periods[2]—as long as the environmental parameters pressure and temperature of the rock do not change significantly—one of the reasons why we keep coming back to the dynamics of our Earth's crust in this book.
- The initially recorded direction is the most stable magnetic recording of a rock. Later directions of the Earth's magnetic field can overlay the original signal, but not destroy it[3]. Laboratory experiments on rock samples are necessary to understand the respective generations of magnetization in a rock.
- The strength of the Earth's magnetic field is, by human standards, low. Accordingly, the magnetization of rocks is also weak. Sensitive, so-called magnetometers are necessary to measure rocks. Another reason why magnetic samples usually have to be brought from the field into a laboratory.
- Since rocks are formed in various ways, the recording mechanisms of magnetization in the respective rocks are also different. For this purpose (not only for this purpose), we distinguish between magmatic, sedimentary, and metamorphic rocks and assign them typical acquisition mechanisms (Sect. 3.2).
- The "freezing" of the magnetic direction in a rock occurs spontaneously, although the process of rock formation can take many thousands of years. An understanding of the geological processes in individual cases is thus a prerequisite for successful interpretation in paleomagnetism and rock magnetism (Chap. 4 and 5).

So let's take a closer look. As already mentioned, only a few of the over 4000 existing minerals on Earth have the ability to store magnetization over periods of millions of years or to record this magnetization—here of the Earth's magnetic field—in the first place at all. Even within the group of remanent magnetic minerals[4] and their ability to carry a magnetization, there are also significant differences. In addition to the strength of the signal, there are particularly differences with regard to the capability of such

---

[2] We will have to briefly discuss different types of rock formation.
[3] There will be several more explanations and analogies on the keyword "vectors".
[4] As a technical term in the sense of the Latin *remanere* ("to remain"), but not necessarily *permanent* ("lasting"). Sect. 3.3 discusses that magnetization, even if it can still be detected after millions of years, does not last "forever".

3 Magnetism

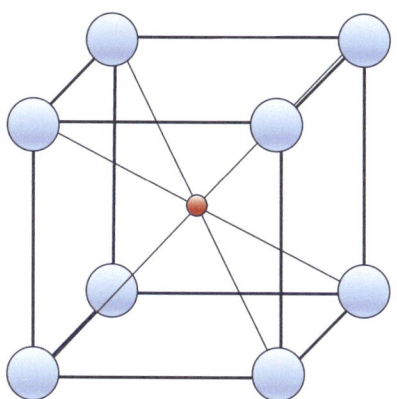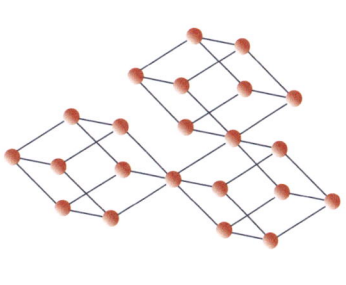

Fig. 3.1 Highly schematic representation for a better understanding of the term "crystal lattice". Elements such as iron or oxygen are arranged in such three-dimensional lattices in minerals. A unit cell can be found as the smallest unit, which is repeated over and over again in such a lattice. In a simple (cubic) arrangement of a unit cell, elements may occupy the corners of a cube. In increasingly complex symmetries (triclinic), one might imagine the unit cell as a building block without any right angles. Different minerals incorporate the most diverse elements in the most diverse arrangements. Combinations are possible, and sometimes areas of the lattice have to be mirrored and rotated to recognize symmetries.

minerals to store multiple magnetization directions, should the Earth's magnetic field change during millions of years since rock formation.

The differing remanent magnetic behavior of minerals is largely explained by the internal structure of the respective minerals. Most minerals—like most other terrestrial matter—are known to be composed of chemical elements at the atomic level. In minerals, these chemical elements are arranged in ordered structures, following certain symmetries (Fig. 3.1) in so-called crystal lattices. Quartz crystals, for example, consist of silicon (Si) and oxygen (O). When cooling from a melt, these two elements arrange themselves in a self-similar, three-dimensional structure. The specific arrangement, possibly even an incomplete lattice, and always the interaction of the chemical elements involved in the lattice play a decisive role for a possibly resulting magnetic remanence of a certain mineral. The book began with the promise not to demand mathematical equations or physical, even quantum mechanical or other scientific understanding of a higher form (although hopefully to arouse interest in it or to rekindle the spark). So let's go.

To understand the causes of magnetic properties of minerals at the smallest, subatomic level, let's first look into a world where we could recognize thousandths of a nanometer with good resolution. This would indeed reveal an astonishing variety of movement in seemingly rigid matter like rocks. For this purpose, let's stick to the idea that the building blocks of atoms behave like ordinary spheres in the more familiar world. An atom consists of an atomic nucleus, surrounded by an electron shell. To help with visualizations of electrons and atomic nuclei, let's briefly dive into a footnote.[5]

A crucial point to understand magnetization is the complex movement of (negatively charged) electrons around the respective (positively charged) atomic nucleus of an element. In a simplified conception, an electron moves on an orbit (shells) around the atomic nucleus and, in addition to this movement on the orbit, also rotates around its own axis (spin). Much like a spinning top on a large circular path or, on a completely different scale, the Earth around the Sun. These two simultaneous types of electron movement (orbital plus spin) generate a so-called *magnetic moment*, as the smallest component. No perfume in this case (Chap. 1), but ultimately *magnetism* in our daily observation world. Different types of magnetism, in turn, are based on the different interaction of the different magnetic moments with each other.

Elements like oxygen atoms and silicon atoms, for example, are built differently. The respective atomic nucleus is larger or smaller (heavier or lighter) and is also accompanied by a different number of electrons. Since the type of electron movement differs from element to element, the respective resulting magnetic properties also differ. Depending on how and which different elements are arranged in a lattice[6], the constantly moving electrons of neighboring elements (better cations) will also influence each other[7].

This is important because elements on Earth usually do not occur in isolation, but in combination with other elements of the Earth's crust as minerals and rocks. Particularly important for paleomagnetism is iron (Fe) from the chemical group of transition metals in interaction with oxygen anions.[8] In other words: "Magnetic" minerals mainly consist of iron and oxygen.

No, we do not want to delve into the depths of magnetic matter after all; there is appropriate space for this in relevant scientific publications if interested. A literature

---

[5] A sheet of paper is about 1 million atoms thick. The atomic *nucleus*, in turn, only makes up about 1/10,000 of a respective atom. The rest and thus the largest space of an atom is taken up by the electron *shell*. This shell is the area where the electrons of an atom "orbit" their atomic nucleus. To get a sense of the scale, imagine a person representing an atomic nucleus. This person could radially walk about 5 km in each direction to reach the outer areas of their own electron shell. 1 million such neighboring people with their extensive "premises" would be needed to represent the paper thickness in the order of magnitude of the atoms. The Earth could, by analogy, be wrapped around the equator about 250 times with these neighboring "properties". After all, the Earth has a circumference of over 40,000 km.

[6] For our considerations, as said, we stick to matter (minerals) whose internal structure at the atomic level resembles a three-dimensional crystal lattice (Fig. 3.1). Some climbing equipment on children's playgrounds could be inspired by such lattices.

[7] The corresponding technical term is *superexchange interaction*.

[8] Goodenough, J.B.,1966. Covalency Criterion for Localized vs Collective Electrons in Oxides with the Perovskite Structure. Journal of Applied Physics, 37, 1415–1421. Important for paleomagnetism are iron oxides in the form Fe_xO_y, sometimes also iron oxyhydroxides in the form FeOOH. Sometimes it is also sulfur (S) instead of oxygen, so Fe_xS_y.

search under the above-mentioned keywords or more generally under entries about *magnetic properties of rocks and minerals* will provide numerous sources and discussions here.

Within the scope of this book, the simplifying idea of a crystal lattice made of solid atomic nuclei surrounded by movable electrons can provide at least two images of their magnetic behavior: Firstly, such an image serves to understand the question of how the Earth's magnetic field can more or less permanently change the magnetization of a rock, but does not cause any macroscopically visible alteration in the rock. In fact, the solid composition of a mineral is never changed during magnetization, for example, the recording of the Earth's magnetic field in geological time.

Let's take a sedimentary rock as an example: A sandstone is made up of individual "sand grains" solidified by jamming or binding agents[9]. Any magnetization of the rock, however it occurs, will not cause any macroscopically visible internal or external change in the arrangement of the grains to each other, for example, a rotation of individual grains. A mechanical rotation of the grains to adapt to a new direction of magnetization does not occur. An external magnetic field influences the grains (so to speak) *internally* by influencing the movements of the electrons—either permanently or temporarily. As if the wind direction changes when grass (electrons) moves on which boulders (atomic nuclei) are lying scattered around, magnetization only occurs in the area of electron movement. We understand the grass image, even though a crystal lattice is obviously three-dimensional.

To the second image: Let's remember the dependence of the number and thus the movement of the electrons on the type and structure of the respective atomic nucleus of a specific element (chemistry book). Depending on which elements are positioned in a crystal lattice in which neighborhood of other elements, the respective electrons will also influence each other differently in their movement in a lattice. However, only very few of the configurations occurring in nature result in a *remanent* magnetization from such superposition and interaction of the electrons. In the first instance, it plays a role whether the electrons of individual neighboring atoms in a crystal lattice interact in a certain way to be remanently magnetic or not.

From everyday experience in dealing with magnets, one can suspect that the already mentioned iron (Fe) is a component of a typical magnetic mineral. Native iron is attracted by a magnet—most of us have certainly experienced such behavior in everyday life.

In addition, iron is one of the most common chemical elements on Earth. Iron ranks alongside elements such as oxygen (O) or silicon (Si) in one of the top four places on the scale of natural chemical elements on Earth. Different rankings in frequency only

---

[9] Loose sand becomes a sandstone in this sense because the individual grains are either compressed so strongly that they jam, or because the grains are cemented together, for example because calcareous waters have successively clogged the spaces between the grains.

result depending on whether the entire Earth or respective sub-areas such as the Earth's crust are used as a calculation basis. However, iron in the Earth's crust—paleo- and rock magnetism mainly deals with rocks from this area—rarely occurs in elemental form (Fe), but almost always combined with other elements as a component of minerals and rocks. Often together with oxygen in so-called iron oxides or also with sulfur (iron sulfides).

And this is a very important aspect for our line of thought, because, depending on how and with which other elements iron is incorporated into a crystal lattice in its divalent or trivalent oxidation states, the magnetic properties of this mineral are correspondingly different. Magnetic properties of Fe-minerals range from "still magnetic" to "completely unsuitable" for storing the EMF. Let's pick out three examples:

1. In one of the combinations common in the Earth's crust, for example, iron and oxygen form the mineral *magnetite* (Fe_3O_4). Without going into the chemical intricacies of magnetite at this point, iron and oxygen in this form represent perhaps the most important recorder of magnetic information in rocks of the Earth's crust. The magnetic moments are arranged very effectively to each other.
2. Another arrangement of iron in a crystal lattice of, a then differently named and constructed mineral like *ilmenite* ($FeTiO_3$) will hardly arouse interest in the search for records of the EMF in rocks. The iron mineral ilmenite has no remanent magnetic properties.[10]

Hematite (Fe_2O_3), on the other hand, also shows stable remanent magnetic properties on its own. Like magnetite, hematite is a combination of iron and oxygen, but these elements are arranged differently in the crystal lattice. For example, unlike magnetite, only trivalent iron is incorporated into hematite, resulting in a changed symmetry of the crystal lattice, among other things.

Like most of the proverbial thousands of different minerals of the Earth's crust, including those without iron in the crystal lattice, ilmenite will also react measurably to a magnetic field[11]. But as soon as the external magnetic field disappears, any magnetic reaction to it in this mineral also disappears. A recording of an external magnetic field and even over geological periods of millions of years does not take place.

If a ball is rolled up a hill but let go before it reaches the top, it rolls back to its starting position. However, if the ball is rolled over the top of the hill and then let go, the ball lands in the next valley, i.e. permanently in a different place than its starting position—as long as it is not rolled away again. A permanent change in the position of the ball would be comparable to the recording of a magnetic field.

---

[10] In combination, however, finest laminae of hematite-ilmenite may be extremely stable recorders of the EMF due to their (electron) interaction, e.g. Kasama et al., 2004, Effects of nanoscale exsolution in hematite–ilmenite on the acquisition of stable natural remanent magnetization, *Earth and Planetary Science Letters*, Vol. 224, Issue 3–4, 461–475.

[11] Such magnetic fields are generated by some measuring devices, for example (Fig. 5.21).

In minerals without remanent magnetic properties, on the other hand, it is impossible to ever roll the ball over a hill; the ascent is steep and long. As soon as no magnetic field is acting anymore (no one is actively holding or rolling the ball), the ball rolls back to the starting position.[12]

The explanation for the magnetic properties of rocks and minerals is therefore to be found on the smallest, at least subatomic scale, in the type of movements of the electrons around their atomic nucleus, as well as their possible mutual interaction. Whether and how this behavior of the electrons is influenced by an external magnetic field or also the neighborhood of the different chemical elements themselves depends in the case of minerals and rocks on the respective arrangement of the building blocks at the atomic level in the crystal lattice.

Outside of the remanent magnetic minerals, quartz, for example, also reacts magnetically differently than mica. After all, these two minerals also have a different "internal structure". Typically, quartz is made up only of silicon and oxygen atoms, while in a muscovite (mica), for example, at least potassium, aluminum, fluorine, and hydrogen must also be accommodated in a regulated manner.

A basic subdivision of matter (here minerals) based on magnetic properties, as shown in Fig. 3.2, is intended to serve our understanding of the magnetic properties of rocks. We already suspect when looking at Fig. 3.2 that the vast majority of minerals, which make up the rocks of the Earth's crust, are not magnetic in the classical sense. The subdivision into diamagnetic, paramagnetic, and ferromagnetic substances will be explained in the following.

3.1 Different Types of Magnetism

Ferromagnetics—Minerals with and without Memory

"Small magnets" fall exclusively into the group of so-called ferromagnetics (right tab in Fig. 3.2) or ferromagnetic minerals. We want to understand the term "ferromagnetics" here simplistically as a collective term for special magnetic properties. In this broader sense (s.l., *sensu lato*), physically speaking, in addition to the ferromagnetics in the strict sense (s.s., *sensu stricto*) due to their specific magnetic characteristics, other groups—for example *ferri*magnetics and *anti*ferromagnetics—are to be distinguished. The elephant on the tab symbolizes, in a figurative sense, the magnetic "weight" of this group, but at this point it should not imply any relationships between magnetism and fauna.

---

[12] In all of this, we ignore for the sake of the simplifying image that a ball rolled up to a height and then let go would of course shoot beyond its original starting position due to the additional potential energy.

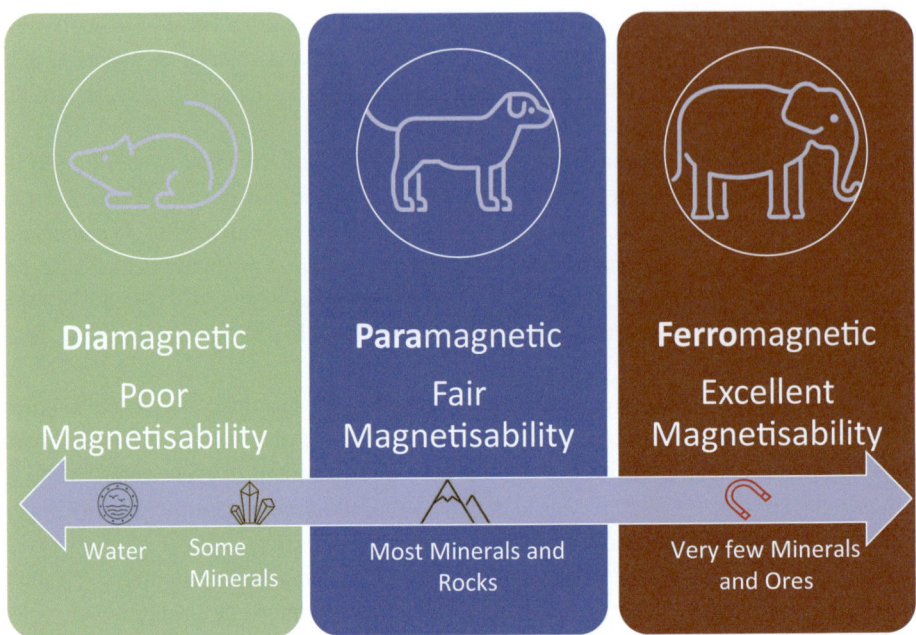

Fig. 3.2 All minerals and rocks have magnetic properties of one kind or another. However, only a few minerals correspond in the classical sense to the common conception of a magnet. Such minerals can record the Earth's magnetic field and remain influenced by this magnetization over geological periods. The majority of minerals as building blocks of rocks only react magnetically as long as an external magnetic field is acting

Only ferromagnetics (s.l.) are minerals relevant for the recording of the Earth's magnetic field. An external magnetic field influences such minerals sustainably. An external magnetic field will strive to magnetically change a mineral in such a way that all electrons—figuratively speaking—follow the direction corresponding to that of the external magnetic field. If a ferromagnetic had no previous contact with a magnetic field, for example immediately after crystallizing, cooling from a melt, this works quite well. When the external field no longer acts, at least some of the electrons remain "aligned" in this way. The memory (physically speaking, one says *hysteresis*) remains and can only be changed again through effort and energy.

In the physical world, these would be the effects of very high temperatures (a few hundred degrees Celsius) or strong electric fields. A strong electric field in nature is conceivable as a lightning strike.

Whether the recording is effective, permanent, and reliable depends on a number of physical criteria specific to different minerals. Among other things, minerals tend to form so-called mixed series in relation to the elements they are made of during their formation. For example, titanium and iron are incorporated into some minerals depending on availability, and physical and chemical criteria, when these minerals form from a rock melt. Iron and titanium have similar properties in some ways (for example, similar size) and can therefore sometimes replace each other as building blocks of a crystal lattice.

Magnetite (Fe_3O_4) is usually an ideal recorder of the Earth's magnetic field at the Earth's surface and then contains only iron, but no titanium. However, if too large a proportion of the places in the crystal lattice were occupied by titanium instead of iron during crystal growth, for example for reasons of availability, this mineral (titanomagnetite) loses the *remanent* magnetic properties, i.e. also the ability to record the prevailing Earth's magnetic field and to preserve this record over geological periods—at least within the range of environmental parameters (temperature, pressure) which we can usually expect in the Earth's crust and at the Earth's surface and which are therefore relevant for most geoscientific questions.

It should be noted here that in addition to the construction of the crystal lattice for the magnetic properties of minerals and rocks, three other factors play an important role: 1) at what temperature and 2) over what period of time we consider magnetization and 3) how large the particle is that is being studied. In fact, a grain of magnetite reacts magnetically differently if it is, for example, 1/1000 or 1/100 mm in size. Most of the magnetic particles in rocks are in this order of magnitude.

An example of what is meant by different magnetic reactions at different grain sizes is the different stability of a once acquired magnetization, i.e. the question of how long a magnetization remains without the influence of *external* factors. External influence in this sense means, for example, that rock layers in which a past state of the Earth's magnetic field was recorded are now included in a mountain building process in a changed geological situation—the magnetic record would thus be exposed to very high pressures and temperatures and thus be forcibly changed or completely erased. Just as described above as a change through "effort and energy". In Sect. 3.3 on relaxation time and especially with regard to the different size of magnetic particles, we discuss the property of magnetic recording, also to decay *without* the influence of *external forces*. The corresponding periods range between seconds and billions of years.

In terms of geological questions, the magnetic properties of a rock desirable for geoscientific investigations are easy to define. The magnetization should hold over geological periods of time, and at usual temperatures in geologically quieter situations of the Earth's crust. Suitable recorders preserve their good magnetic properties up to ambient temperatures of a few hundred degrees Celsius and thus up to a temperature which is only exceeded in the upper areas of the Earth's crust in rare geological scenarios.

Purely physically speaking, some minerals can also record a magnetic field outside the mentioned geological framework. However, if such a magnetic record only occurs at temperatures below 0 °C or over periods of only a few milliseconds, it will have no relevance for the considerations of magnetization in this book. And thus, in the rarest cases, it will be suitable for practical application with a geological background.

Purely scientifically speaking, all magnetic properties of all minerals in all states are studied. Those who want to delve deeper into the subject will learn to understand more about the relationships through the keywords "relaxation time", "domain status" and "Curie temperature".

In rock magnetic studies, the physical quantity of *magnetic susceptibility* is often considered. It is a measure of the magnetizability of a rock in response to an external magnetic field or, in other words, susceptibility acts as a kind of multiplication factor for different materials (here minerals), which leads to a more or less strong "magnetic reaction" of the respective material to the same external field. In Sect. 3.5, we discuss the magnetic susceptibility of minerals and rocks in more detail. At this point, we only want to note that the strength of the magnetic susceptibility does not determine whether a mineral can store the Earth's magnetic field, i.e., whether the mineral has remanent magnetic properties.

The "reaction" to an external magnetic field is strongest for ferromagnetics compared to other materials outside this group. Within the ferromagnetic minerals, a range of individual magnetic susceptibilities is covered at this comparatively high level. Magnetite, hematite, cobalt, and iron belong to this group.

In geosciences, specifically in rock magnetism, the distribution and frequency of individual ferromagnetic minerals in the Earth's crust are crucial factors for successfully answering many research questions. For magnetic studies, it is advantageous that minerals such as magnetite are contained in almost *all*[13] rocks of the Earth's crust, at least in traces. The volume fraction is often so small that in relation *one* grain of magnetite builds up the rock together with 10,000 other components, such as quartz or feldspar.

It may initially seem surprising in this sense that extensive studies are carried out in paleomagnetism on rather unmagnetic chalk cliffs. However, due to the strong magnetic signal, even less than 1/100% content of finest magnetite justifies and rewards these efforts, at least since the development of sufficiently sensitive measuring devices[14].

In field studies and especially in the laboratory, the magnetic signal of even the smallest traces of magnetic minerals can be detected and analyzed. Also in magnetically weak, but especially widespread carbonate rocks along the world's coasts. Due to the dynamics of the Earth's crust already mentioned, these weakly magnetized rocks contain information about the Earth's magnetic field of the more recent Earth history—let's say, the last 200 million years—(Chap. 2).

In contrast to widespread mineralogical "fixtures" like magnetite, pure iron or cobalt are less prominent in paleo- and rock magnetic studies due to their low distribution in

---

[13] This includes magmatic rocks such as ocean floor basalts (Chap. 2), but also hardly magnetic sedimentary rocks, for example salt rocks, which are formed in restricted lagoons by the evaporation of seawater.

[14] Here we are referring in particular to superconducting sensors of so-called cryogenic magnetometers, which are cooled down to near absolute zero at -273 °C using liquid helium. Such magnetometers have increasingly expanded the range of magnetometers used in paleomagnetic research since the early 1990s, thus making enormous amounts of—until then unused—magnetic information in rocks accessible for research.

3.1 Different Types of Magnetism

the Earth's crust. The occurrence is limited to a few ore deposits and the occasional individual find, so no extensive spatial or temporal coverage in geological terms can be expected.

A high magnetic susceptibility does not necessarily guarantee an effective recording of the Earth's magnetic field in the geological past. A clear reaction of a rock to a weak external magnetic field does not mean that this reaction is also permanently (remanently) recorded. In fact, only very few minerals have the magnetic properties to effectively record and store the Earth's magnetic field. There are only about ten such remanently magnetic minerals among the thousands of known minerals on the Earth's surface and in the Earth's crust. We will later explain which magnetic applications are possible even without the recording of the Earth's magnetic field. When we use magnetic susceptibility to, for example, distinguish individual rock layers from surrounding rocks, the clear magnetic signal of the ferromagnetics also helps us—with or without recording of the Earth's magnetic field.

Paramagnetics—never remember

The middle tab in Fig. 3.2 names paramagnetics. In the sense of the explanations for the group of ferromagnetics, these minerals are characterized by a medium magnetic susceptibility. Paramagnetics include a diverse group of the most common minerals and rocks of the Earth's crust, all of which are never suitable for storing the Earth's magnetic field. The latter is at least the apparent criterion for work in paleomagnetism.

An external magnetic field, such as the Earth's magnetic field or a magnetometer, causes a magnetic influence on paramagnetic minerals—as explained above, the external magnetic field influences the movement of the electrons at the atomic level. The stronger the acting magnetic field, the stronger the reaction will be. In contrast to remanent magnetic substances, this reaction disappears linearly with a decaying external magnetic field. For susceptibility measurements, paramagnetics are an important part of the rocks due to their wide distribution. Olivine, biotite (a mica), and siderite belong to this group.

Diamagnetics—even water is magnetic

The left tab in Fig. 3.2 names diamagnetics. The expected magnetic reaction is low compared to the other materials and is actually even less than zero. A magnetic susceptibility less than zero, i.e., a negative susceptibility, sounds strange at first. In a – limping – analogy to the magnetic reaction of diamagnetics to an acting magnetic field (usually a measuring device), one might imagine the following: A beam of light from a flashlight is reflected back very weakly; with paramagnetics (see above), the beam of light would be further directed in its direction. Consequently, this means that diamagnetics lower the total value of magnetic susceptibility when measuring a rock sample, while para- and ferromagnetics increase the sum (more on this later in Sect. 3.5). Diamagnetics are also not suitable for storing the Earth's magnetic field. However, since the widely spread quartz in rocks and, for example, water are among the diamagnetics, these substances can also make a significant contribution to magnetic measurements when it comes to rock characterization. Regarding the supposed importance of water in this context, consider that groundwater, pore water of rocks, rock moisture, etc. often have a significant (volume) share in the rocks of the Earth's crust. Now imagine a sandstone made of quartz grains with water-filled pores and the claim to still want to conduct magnetic investigations (more details in Chap. 5).

3.2 Rock Type Determines Acquisition Mechanism

The Earth's magnetic field can be recorded in all types of rock, whether it is magmatic rocks or, for example, sediments deposited in a lake. Usually, the first magnetization, i.e., recording of the Earth's magnetic field, occurs during an early stage of rock formation. In the case of magmatic rocks, this is, for example, the crystallization of minerals from a cooling rock melt. This original magnetization remains preserved under average conditions at the Earth's surface over geological times, unless the rock is heated to very high temperatures (several hundred degrees Celsius), struck by lightning, or, for example, eroded.

On the following pages, we want to take a closer look at some of these magnetic acquisition mechanisms. Some basic properties of acquiring a remanent magnetization will be enough to come a lot closer to understanding. In these considerations, we refer solely to the small group of the aforementioned ferromagnetic, remanently magnetic minerals. As said, the macroscopically visible minerals of a rock, for example, quartz grains of a sandstone, are not capable of recording a magnetic field. The different ways of storing a magnetic field will be explained in the following.

Heat and Cooling

A magmatic rock is formed from a cooling rock melt by successive crystallization of the mineral components—until the rock, for example a granite, is completely solidified. As soon as a mineral-specific temperature of suitable minerals is undercut, the magnetic field in the emerging mineral is "frozen" (Fig. 3.3)—this is called a thermoremanent magnetization (TRM)[15]—, of course not in the literal sense of the word, because such mineral-specific temperatures are still at several hundred degrees Celsius. And of course not in all minerals of our imagined granite, which consists almost exclusively of the minerals feldspar, quartz and mica, none of which is suitable for recording an earth magnetic field.

We may understand the recording of a field in such a way that remanently magnetic minerals contained in the magma orient themselves along the field lines at the location of the lava extrusion. Tiny magnetite particles, for example, remain permanently oriented

---

[15] The classic scientific article on this: L. Néel, 1955. Some theoretical aspects of rock magnetism, *Adv. Phys.*, v. 4, 191–242.

3.2 Rock Type Determines Acquisition Mechanism

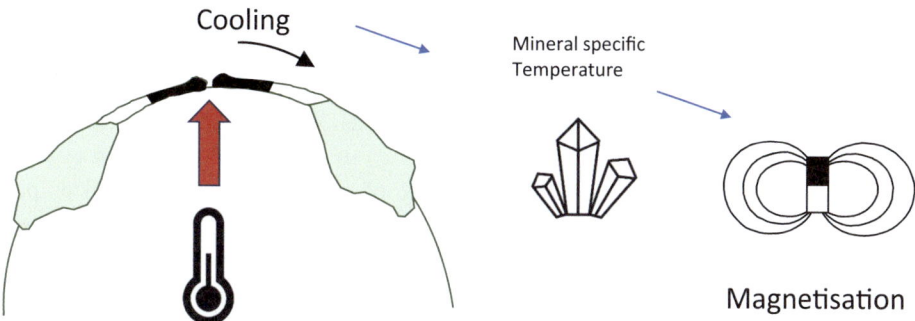

Fig. 3.3 High temperatures prevent a remanent magnetization for physical reasons. Also, actually suitable magnetic minerals have an individual (Curie-)temperature depending on the type of mineral, only below which a long-lasting storage of the earth's magnetic field becomes possible. A possible geological situation is shown schematically, in which magma intrudes and is only suitable for storing a magnetization after a certain cooling. Note, however, that Curie temperatures of prominent magnetic minerals often lie above 500 °C. Cooling is therefore quite relative

in the direction of the prevailing earth's magnetic field below 573 °C (the so-called Curie temperature of magnetite). In Sect. 3.3 we will discuss the term relaxation time[16]. When cooling from a lava, the time during which the magnetic orientation of the "small magnets" can be maintained increases exponentially as the Curie temperature is approached. At 1000 °C at the time of the lava's extrusion, an orientation cannot even be maintained for 1 second; below the Curie temperature, the relaxation time quickly becomes millions of years.

This process of aligning magnetic particles is magnetically effective, especially since magmatic rocks contain a high percentage of remanently magnetic minerals (>5 %) compared to other types of rock. In particular, the magnetization does not require any (mechanical) alignment or rotation of a magnetic particle along the field lines to be recorded. The magnetization takes place internally by influencing the electrons at the atomic level (see above). This influence will occur similarly in simultaneously cooling minerals in a melt. Lava does not cool down in its entire mass at the same time, but from the outside in. Depending on how quickly the cooling takes place, changes in the earth's magnetic field within a magmatic sequence (in the direction of cooling) may well be recorded.

Obviously, lava cools faster at the earth's surface or when flowing out under water than deeper down in the earth's crust, so there can also be differences in the efficiency of the recording here. By efficiency we mean that each individual remanently magnetic mineral optimally represents the earth's magnetic field (e.g. Fig. 2.7).

---

[16] Every remanent magnetization decays. If this process takes millions of years, there is enough time to measure this magnetization in the laboratory, so also in very old rocks. High temperatures may shorten the relaxation time from millions of years to fractions of seconds. For geological studies, this is too short.

Rotating in Pores

After just a few weeks of studying paleomagnetism, the term "post-depositional detrital remanent magnetization" will be a quite ordinary term and will simply conveniently name the magnetic recording mechanism of sediments. At least such sediments, which are collected from the eroded components (detritus) of earlier rocks and then re-solidified into a (sediment) rock—a very typical type of clastic sediment. A small proportion (usually less than 1/100 of the volume or under 0.01 %) of the detritus will consist of tiny, micrometer-sized remanently magnetic minerals. Often these are the aforementioned magnetite, but of course other iron oxides or iron sulfides such as hematite or greigite are also present. Since these magnetite grains were already magnetized in the earlier rock from which they have now been eroded, these components can be imagined as small independent magnets.

For example, if these particles slowly sink through a water column to the bottom of a lake, they will align along the field lines of the Earth's magnetic field. Since these particlesare, as mentioned, already small magnets, they will actually be mechanically rotated into alignment along the field lines. This is quite contrary to the previously outlined internal magnetization of the particles during thermoremanent magnetization (TRM). If the deposition site were, for example, in high northern latitudes, a magnetic particle (for simplification, we can imagine a tiny bar magnet) would align along the local magnetic field lines at a very steep angle against the Earth's surface. Closer to the equator, this angle of inclination becomes steeper, as already described (e.g., Fig. 2.7).

However, it's not quite that simple, because we still have to explain the term "post-depositional"—in the sense of "after deposition": If we imagine a small bar magnet landing on the Earth's surface, it will likely be pushed around with the load of further, even non-magnetic grains. In other words, the crowding of surrounding, contemporaneously deposited sediment grains will push the small bar magnet out of its ideal orientation along the magnetic field line. As a rule, the angle expected according to the field lines (the magnetic inclination) against the Earth's surface becomes flatter—a newly forming sediment layer, and with it the magnetic record, is "flattened" by the increasing load of new grains [17](Fig. 3.4).

---

[17] In paleontology, fossils of past life forms are of course the center of interest. It may be the lithified skeleton of a dinosaur or imprints of marine microfossils, whose delicate traces can only be explored with high-resolution microscopes. However, all fossils have in common that heat and pressure after embedding can make the fossil increasingly unrecognizable, even almost unusable. As an example of the effect of pressure (here as increasing load) in the Earth's crust, imagine the exhibited skeleton of a dinosaur in a natural history museum and what would happen if we were to shovel tons of sand onto the exhibit. Bones would be pushed out of their compound, and the three-dimensional shape would increasingly be flattened—banally speaking, like a moderately inflated ball that one sits on. A sphere becomes an ellipsoid, a circle becomes an ellipse (Fig. 3.4). In a similar way, such a deformation can also happen to the recording of the Earth's magnetic field in rocks—at this point we want to imagine a stored magnetic direction as a three-dimensional structure. We will better understand what such a structure means in the following.

3.2 Rock Type Determines Acquisition Mechanism

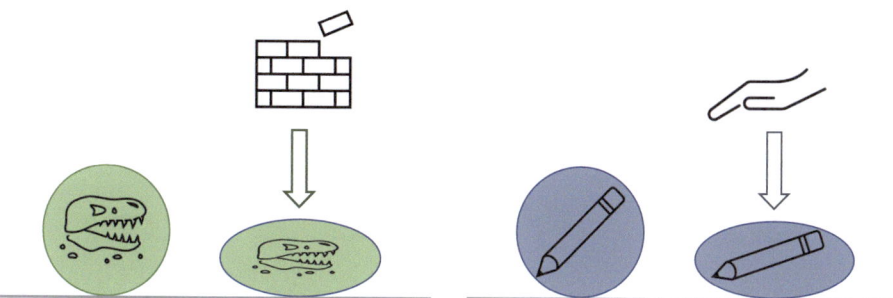

Fig. 3.4 Just as directed pressure in the Earth's crust, for example through load accumulated over millions of years of sediments, deforms the three-dimensional shape of an embedded dinosaur skeleton, the magnetic information stored in rocks about past states of the Earth's magnetic field can also be subsequently deformed. Especially information about the directions of the Earth's magnetic field. The deformation of a stored magnetization direction in rocks can be imagined analogously to a pencil (any elongated object equally) in a block of foam. If the block is flattened, the pencil also rotates into a flatter position. Both in paleontology and in paleomagnetism, such subsequent changes of the "fossils" must be taken into account and corrected in order to derive correct information about the geological past of the Earth

However, we have overlooked a crucial aspect in this picture. Magnetic particles relevant for recording are small, especially compared to simultaneously deposited non-magnetic particles, which make up the main component of the newly formed sediment (Fig. 3.5). Let's assume that quartz grains are the main components of a newly formed sandstone. Let's assume the size of the quartz grains of a fine sandstone to be about 0.1 mm. A suitable magnetite particle is only about 1/1000 of this size. A striking comparison: If the magnetite particle were the size of a tennis ball, the length of the quartz grains in our example would correspond to the height of the cathedral towers in Cologne. This example is meant to aid our imagination; of course, there are both finer and coarser sediments in nature. In the case of the finest clays, the size difference between the main components and the magnetic particles may almost disappear, but then the balance of forces between magnetic alignment and mechanical "jostling" also shifts.

Between the quartz grains, as in our example, unfilled gaps remain, as the grains do not fit together ideally—the pore space (Fig. 3.6). From our striking comparison with the cathedral towers, it becomes clear that the magnetite particles have enough space to move freely in the pores between the other components. Especially when this pore space is filled with water (let's stick to our example of deposition on a lake bed), the "small bar magnet" can freely align itself along the field lines of the Earth's magnetic field even long after deposition. This process may be more effective in terms of optimal alignment parallel to the field lines of the total magnetite particles deposited at a time than the original sinking of the particles through the water column of the lake. After all, other forces are at play in the latter process, such as the slightest currents in the water, which may affect the alignment of the particles. The particles aligned along the field lines in the

Fig. 3.5 The ideal recording of the inclination of the Earth's magnetic field could be influenced in sediments, at least by the space available in an association of neighboring grains. In ① a magnetic, elongated particle is shown, which aligns itself along the field lines of the prevailing Earth's magnetic field. This could happen, for example, when the particle sinks in the water column of a lake. In ② it is shown how this particle is pushed out of its orientation by the crowding of other deposited sediment grains. However, magnetic particles are often so small in relation to the other particles of the sediment that they can rotate into their ideal position along the field lines in the pore spaces between the non-magnetic grains even after deposition (Fig. 3.6)

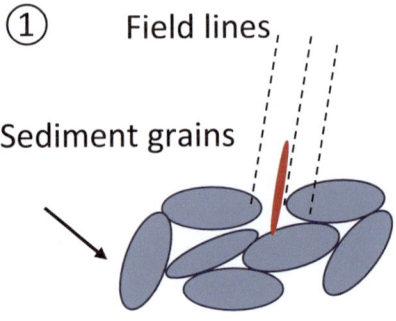

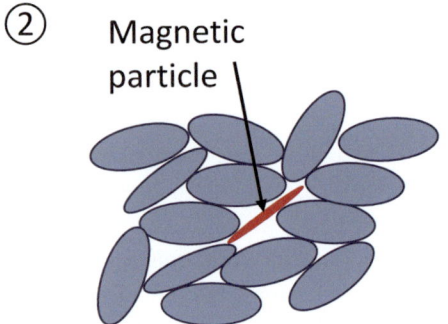

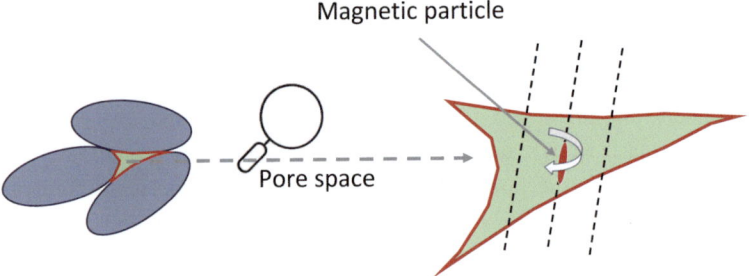

Fig. 3.6 Often, magnetic particles that record the Earth's magnetic field in sediments are so small in relation to the other sediment grains that they can comfortably rotate in the pore space between these grains. There, the particles will be able to align themselves ideally along the field lines of the Earth's magnetic field, unaffected by the crowding outside the pore space. The illustration shows schematically an enlarged view of such a pore space between non-magnetic grains. A depositional magnetization in sediments often occurs with a short delay after the initial arrival of the particles on the sediment surface

3.2 Rock Type Determines Acquisition Mechanism

pore space are "fixed", and the Earth's magnetic field is thus recorded as soon as the pore space is drained in the course of the progressive lithification of the newly formed sediment and further movement is thus prevented.

It is obvious that with a post-depositional detrital remanent magnetization, there may be a temporal gap between the time of the first arrival of the sediment particles on the (lake) surface and the recording of the Earth's magnetic field. After all, it would not be expected that drainage and increasing lithification would occur directly or at least too close to the sediment surface. For this process, more and more particles must be sedimented onto the surface.

The length of this period until the storage of the magnetic signal is a function of many processes that influence the "fossilization" of the sediment after deposition. This includes, for example, the rate at which sediment particles are deposited, in other words, how quickly a deposited particle is increasingly buried by subsequent particles. But also the prevailing shape and size of the introduced particles and thus especially the environmental conditions of the depositional environment are important. It makes a difference whether finest particles are delivered as suspended load in the middle of a lake or whether, for example, a mountain stream with high transport energy carries correspondingly large pebbles.

Upon closer examination, one must also ask whether the newly deposited sediment was churned by organisms[18] until it solidified. Obviously, this would prevent any alignment of magnetic particles. Storage could only occur below the churned zone, where no mechanical disturbance occurs anymore.

Another problem, depending on the deposition space, may be the decomposition of other components of organic material introduced into the upper sediment layers. Such degradation of organic material (i.e., processing for energy gain/as food) is carried out by bacteria (Chap. 5). Such bacteria are adapted to the different chemical conditions in the increasing depth zones of the sediments. Particularly in the upper centimeters, there are both types of bacteria that use existing magnetic particles as food and bacteria that can *produce* and excrete finest magnetite particles.[19] Such excreted magnetite particles also record the prevailing magnetic field. They obviously did not originate simultaneously with the previously detritally introduced magnetite particles and record a possibly different configuration of the Earth's magnetic field (Sect. 3.4).

In our example, we have chosen a lake as the deposition space; we could have chosen river, sea or dune sediments with equal justification. Overall, as is often the case in geological scenarios, it must be decided on a case-by-case basis what influence sedimentary

---

[18] For example, "grazing" worms churn the upper layers of sediment. Other organisms systematically graze the surface. Lithified grazing traces can be found in rocks as ichnofossils, e.g. W. Miller, III (ed.) 2007. *Trace Fossils. Concepts, Problems, Prospects*, Elsevier.

[19] For example, Petersen, N., von Dobeneck, T. & Vali, H., 1986. Fossil bacterial magnetite in deep-sea sediments from the South Atlantic Ocean, *Nature,* 320, 611–615.

processes had on a possible delay in magnetic recordings. If there is a delay between sediment age and magnetic recording, it will usually be less significant in rocks millions of years old. To understand the order of magnitude: for example, the delay between deposition and recording of the Earth's magnetic field in selected lake sediments is about 2000 to 3000 years[20]— it then depends on the question of the investigation to what extent a delay in the recording will influence the result.

Growing, growing

Another way for sediments to record the magnetization of the Earth's magnetic field is the so-called chemoremanence. This does not mean that the magnetization is particularly resistant to later chemical influences, but rather the following: Minerals grow. Starting, for example, from a tiny nucleus in a nutrient solution[21], to which material increasingly accumulates. Remanent magnetic minerals must indeed reach a minimum size in order to "remember" information, i.e., to permanently record the Earth's magnetic field (Fig. 3.7).

The preceding sentence contains two statements, each of which could fill its own chapter. For understanding the recording of the Earth's magnetic field in rocks in the context of chemoremanent magnetization, the following consideration should therefore help us: Not all sedimentary rocks are made up of detrital components, as most examples in this book might suggest. Rocks that are formed by the evaporation of (sea) water are also counted among the sediments. Salt and carbonate rocks from the Permian period belong to this. The crucial difference is that the precipitation from a supersaturated solution occurs at low temperatures, unlike the solidification of a 1000 °C hot magma. At ambient temperatures, as we are accustomed to on the Earth's surface, magnetic minerals can form in this way, for example on the shore of a lake, a river, or in crevices and on fracture planes of older, tectonically stressed rocks.

Minerals also grow in the case of low temperatures, with more and more material accumulating around an initial nucleus, much like lime deposits accumulate on a faucet. An initial nucleus has a size of only a few thousandths of a millimeter. As soon as a size, which varies depending on the type of mineral being formed but is still in the micrometer range, is exceeded, the newly formed mineral stores the Earth's magnetic field. To use an analogy mentioned earlier: The magnetization is stored, much like small branches of trees are no longer moved by mild wind when these branches have grown into stronger branches.

---

[20] Snowball, I. et al., 2013. An estimate of post-depositional remanent magnetization lock-in depth in organic rich varved lake sediments. *Global and Planetary Change,* Vol 110, Part C, 267–277.

[21] As a geological example, this could be seawater evaporating in the sun in the back area of a lagoon. In everyday life, one could observe how a sugar solution crystallizes. In both cases, the individual crystals grow by gradually accumulating more and more material around an initial, tiny core. Of interest to us: "Magnetic" minerals can also grow in this way.

3.3 Size and Time 75

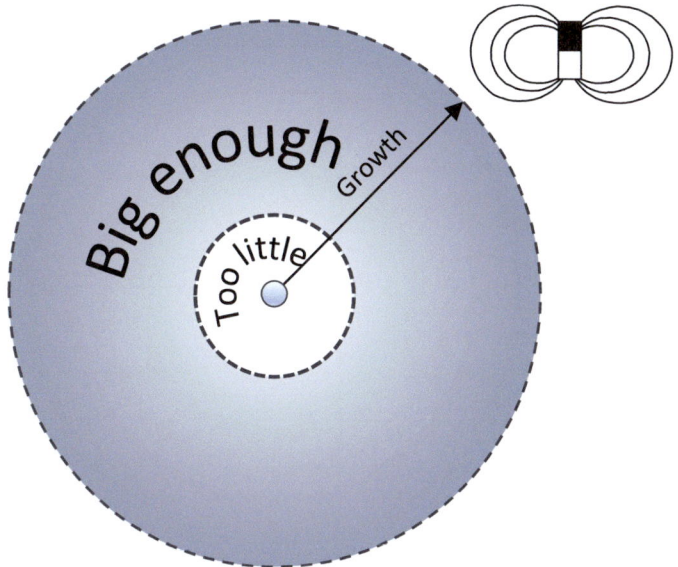

Fig. 3.7 Just as some particles are too hot to store the Earth's magnetic field for a long time, other particles are too small. When particles in a nutrient solution grow around an initial nucleus, they occasionally reach a mineral-specific, sufficient size. For scaling: The particles are still so small that they would hardly be recognizable with the naked eye

3.3 Size and Time

Relaxation Time
From a more scientific perspective, the cited thickness of the branches or the size of magnetic particles is defined as the duration of the storage of magnetic information—the so-called relaxation time—(Fig. 3.8). This time is a function of parameters such as the type and properties of the emerging mineral, its size (more precisely volume) and the ambient temperature at which a mineral grows. The already cited Louis Néel has shown that the listed characteristics are exponentially linked to the relaxation time under certain conditions. In other words, even small changes in one of the determining characteristics can have a dramatic effect.

If the type and properties of the growing mineral remain the same—let's say a magnetite particle is growing and it remains, as expected, a magnetite particle—and the ambient temperature also remains the same, then the changing volume due to growth is the only parameter that changes the resulting relaxation time. In the (magnetically) least favorable case, the constellation of the mentioned input parameters results in a preservation of the magnetic recording of only a few seconds before the magnetic order is destroyed again. Such a state is referred to as superparamagnetic and is defined with a relaxation time of

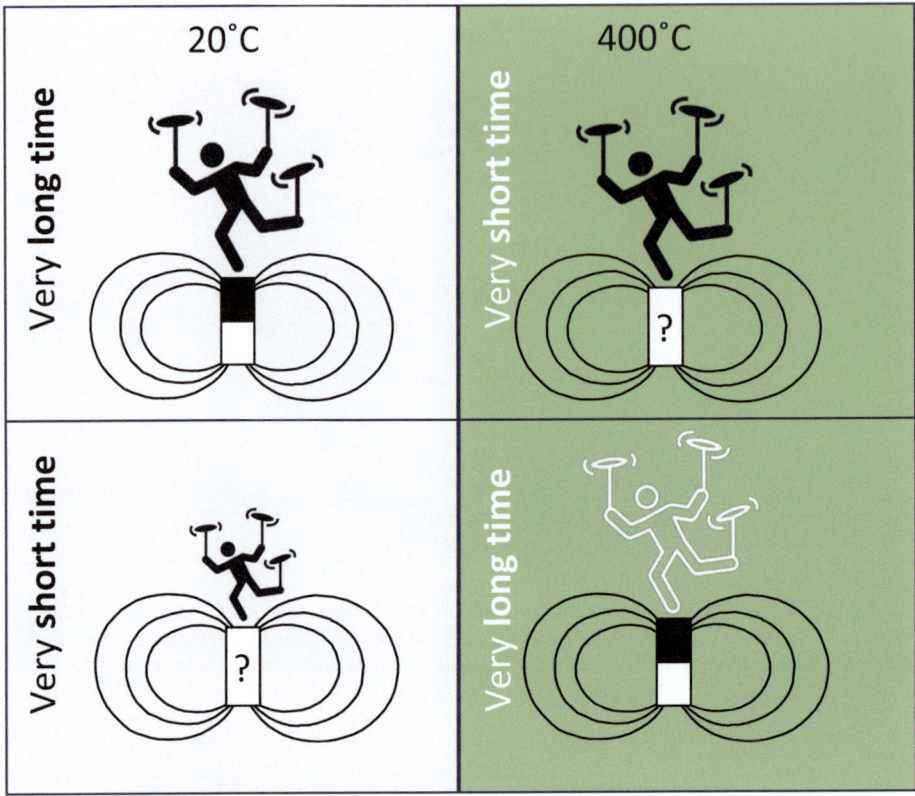

Fig. 3.8 To understand the term "relaxation time". The definition of a remanent magnetic mineral is based on the duration this mineral can preserve a stored magnetization. At least 100 s is commonly considered as the minimum limit; for geological matters, many millions of years are usually required. In addition to the "construction and type" of a mineral, the relaxation time often depends on the size of the particles and the ambient temperature. The left side of the figure illustrates that the same mineral at a typical surface temperature on Earth may be too small to store magnetization for a long time. The right side schematizes that a remanent magnetic mineral at the same size, but increased ambient temperature, loses this magnetic property. Another type of mineral may, despite the same size, hold its magnetization for long periods of time, even at high temperatures (bottom right)

less than 100 s. If, as in our example, the volume of the particle changes only minimally, the relaxation time increases very quickly to geological periods of millions of years due to the exponential linkage of the parameters.

This apparent mix-up of size (volume) and duration (relaxation time) may seem confusing at first glance. Put differently, this definition simply means that, for example, a magnetite particle must reach a different size during its growth to store the Earth's magnetic field than other types of remanent magnetic minerals do. For instance, due to its different crystal structure and chemical composition (see above), hematite must reach a larger volume than magnetite to leave the superparamagnetic state. Expressed through the relaxation time, all magnetic minerals have a uniform limit beyond which a remanent magnetic record becomes possible. For use in paleomagnetic studies, the relaxation time must be longer than the age of the rock formation to be studied. Often, therefore, times of several hundred million years are required. We remind at this point of Chap. 2 and the brief look at the geological timescale there.

The big ones too big, the small ones too small
At the end of these considerations, it should be noted that increasing growth of magnetic particles does not lead to ever longer relaxation times, i.e., the magnetization would be preserved longer. Rather, there is an ideal grain size range for recording the Earth's magnetic field. Such "ideal" particles are small, but not too small. The decay of an acquired magnetization has already reached a duration of billions of years here.

For better appreciation: For magnetite, this size range is between 0.01 and 0.1 µm. Such particles are referred to as single-domain particles. By "domain" we mean so-called magnetic domains, i.e., smallest areas of magnetic uniformity, which can be preserved more energetically favorable than larger areas. Somewhat like it might be easier to heat a small room in the house when needed, rather than constantly keeping an open-plan configuration temperate. Of course, boundary conditions such as house insulation, heat source, and much more must be taken into account. The same applies analogously to magnetic domains in different types of remanent magnetic minerals.

Magnetization is a complicated physical process. In Chap. 2 we have already mentioned the scientific term entropy and the well-known fact that it takes work to maintain order. When a particle "externally" continues to grow, "internally" an order into smaller magnetic areas (the aforementioned domains) occurs for physical/energetic reasons. The larger a magnetic particle becomes, the more such internal domains may arise—always depending on the mineralogy of the particle.

When a magnetic field like the Earth's magnetic field is recorded, all domains align magnetically the same way. For simplicity, let's imagine the respective domains as individual small arrows, which either point in a common direction (magnetized parallel to the field lines of an external field) or "randomly" in all possible directions and thus show no preferred direction.[22] Maintaining an order (i.e., the magnetization) within the domains costs energy (we can all quickly think of an analogy from everyday life). Left to itself, this principle will lead to a lower stability of a once acquired magnetization compared to the stable single-domain particles. Lower stability in this case means that, starting with the domains of larger magnetic particles, increasingly more arrows are turned out of the common preferred direction. In other words, despite their size, a recording of the Earth's magnetic field acquired by multi-domain particles is not more stable than for single-domain particles of ideal size—on the contrary.

---

[22] In Fig. 3.15 we will discuss a similar problem of aligned arrows again. For illustration, you may also use the figure at this point.

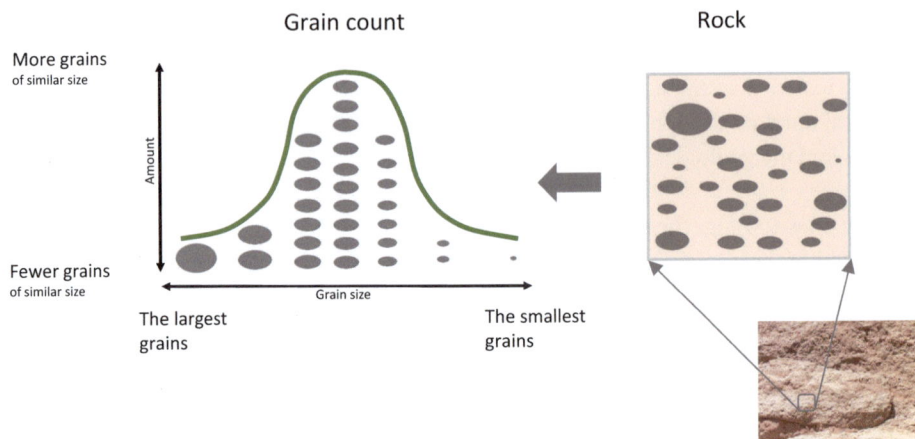

Fig. 3.9 A so-called Gaussian normal distribution of grain sizes shown using the example of a sedimentary rock. The diagram on the left illustrates the also common term "bell curve". The majority of grains are in the middle range of the bell. Deviations from the prevailing grain size decrease uniformly in each direction. We are familiar with variations of such distributions from numerous considerations in everyday life, from the age distribution of a country's population to the representation of financial variations in stock exchange indices. In rocks, the grain size distribution applies to both the entirety of the magnetic and non-magnetic components, but also to the remanently magnetic particles of the rock considered separately. If one were to mentally—or graphically as a representation of underlying measurement data—compress the bell laterally, this would correspondingly represent an increasingly uniform grain size. If one were to flatten the bell or steepen it on only one side, this would indicate an increasingly wide distribution of underlying grain sizes or an underrepresentation of certain grain sizes. Such deviations from an average bell can be explained, for example, by subsequent changes to the deposited material. For example, secondary precipitation in the pore spaces of a sandstone could create a generation of grains completely independent of the original deposition process. Both generations may each correspond to a well-formed bell, but mixed together in a sample they may produce a "deformed" normal distribution. Geological processes in the realm of rock formation represent a multitude of possible variants—beyond this book

The above considerations regarding the size of magnetic particles apply regardless of the respective formation process (magmatic, diagenetic, sedimentary) of the minerals. In particular, it should be noted in relation to the application of magnetic methods in the geosciences that all rocks have a *distribution of grain sizes* (magnetic and non-magnetic) of minerals. In other words, not all particles have exactly the same size, but they are subject to a certain variation (Fig. 3.9). If one were to sort the components of a rock by size, a dominant size would be found due to the number. But there would also be smaller quantities of both increasingly smaller and larger grains. Some geological processes, under which magnetic rock particles are formed, cause a greater range (wider distribution) than other processes. As a result, some rocks may be better or less suitable for some magnetic applications—or at least require more (laboratory) work to reveal the correct magnetization. More on this later.

3.3 Size and Time

Important here: A grain size distribution of one kind or another is to be expected for both magnetic and non-magnetic components of a rock/sample, both in magmatic, sedimentary or even metamorphic rocks. Since the grain size of remanently magnetic minerals has an influence on the stability of a magnetic recording, their consideration including an analysis environmental factors controlling the respective grain size distribution, is always an essential part of a successful paleomagnetic investigation.

Another important aspect, however, arises from the fact that one specific geological process in the course of a rock's formation will usually lead to a normal distribution of grain sizes. Subsequent stress on this rock—for example, a deformation during tectonic activities—which also changes the mineralogy, can lead to a completely new population of magnetic minerals. The old and the new population of minerals can then be visualized as two overlapping bell curves. One possible route of data analysis will then try to reconstruct these different processes in the formation of the rock (Fig. 3.10).

In Chap. 4 we want to deal with the possibilities of using the recording of the Earth's magnetic field for dating rocks. At least one distinction will have to be made here with regard to the quality of the recordings of the Earth's magnetic field in different types of rock (and let us understand why we reconstruct a journey between New York and Cologne in Chap. 4): Depending on the acquisition mechanism, it is more or less likely that a wider range of particle sizes will occur within a rock layer. For example, remanent magnetic particles will have a narrow range of sizes if they precipitate from a "nutrient solution", i.e., for example, when some carbonates are formed. On the other hand, if other remanent magnetic minerals are washed together as detrital particles in a sedimentary process, we expect significantly different grain sizes in the resulting rock.

In general, it can be said in this context that with the increase in the bandwidth of the grain size of magnetic minerals, the likelihood of the recording of the Earth's magnetic field at different times in Earth's history has occurred. The reason is: If different grain

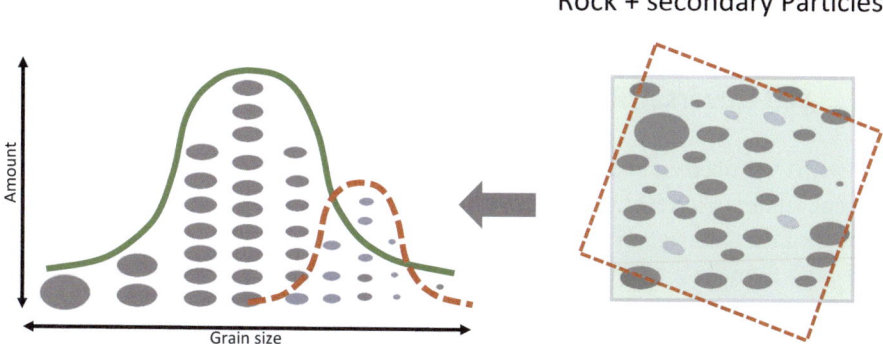

Fig. 3.10 Secondary geological processes after the original rock formation can lead to different generations of grains with different grain size distributions in a rock. Particular caution is required with magnetic minerals, as the same type of minerals with the same particle sizes, but different ages, may have recorded different states of the Earth's magnetic field (Chap. 4)

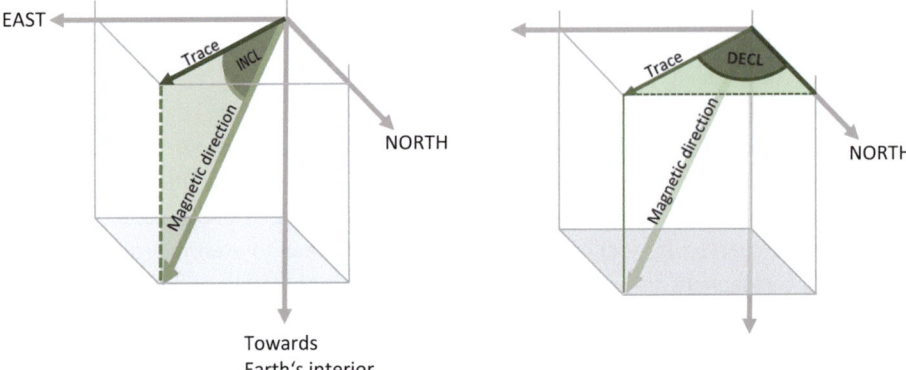

Fig. 3.11 Two important features in defining a magnetic direction in paleomagnetism are inclination (left) and declination (right). The declination (DECL) refers to the angle between the true north direction (in the sense of geographical north or also the point where the Earth's axis of rotation pokes out; Fig. 2.7) and the measured magnetic direction in a sample. For further understanding, it is important that the angle of declination is measured, so to speak, in a plane parallel to the Earth's surface. This happens regardless of the angle of inclination (INCL) at which the magnetic direction dives into or looks out of the Earth's surface. The *trace* of the magnetic direction—as a projection into the horizontal plane—illustrates this. As an example from everyday life, imagine a top-down view of a line diving downwards

sizes of remanent magnetic minerals are more or less magnetically stable, they may be more easily altered or at least influenced by later Earth's magnetic fields.

If you take a rock sample to perform rock magnetic measurements, there is always the possibility that different generations of magnetic recordings of the Earth's magnetic field are *simultaneously* contained in one sample. An essential step in paleomagnetic investigations is to distinguish the individual recordings of the Earth's magnetic field from different times in Earth's history in the same rock sample. And then to interpret them correctly. In the following, we will therefore expand our deliberations a bit further.

3.4 Self-Reversal, Reference Systems and Arrows

Now we have gained an insight into the more physical basics to better understand paleomagnetic studies. However, to understand Chap. 4 and 5 and thus especially the geological applications of magnetic investigations, we still need to discuss a few more things.

In order to make a relevant statement for the applications mentioned, the paleomagnetic investigations must be scientifically sound and thus comprehensible. This would not be possible without investigating and scientifically establishing the physical basics of the magnetization of rocks—down to the smallest level, into the world of quantum

3.4 Self-Reversal, Reference Systems and Arrows

mechanics[23]. It would not be possible to work reliably with paleomagnetic information without understanding the causes of the generation of the Earth's magnetic field itself or coming as close as possible to a complete understanding of them. Paleomagnetic investigations would not be understandable without knowledge of the highly variable temporal and spatial properties of today's current magnetic field of the Earth.

General statements like the above are often better understood in the right context. So let's digress even further.

Let's assume we have examined a series of similar rock samples and found that some of the samples provide deviating magnetic results. Let's assume we are in the 1920s and we already understand a lot about the magnetization of rocks, but not everything yet.

Let's further assume that we have tried to infer the strength of the magnetizing former Earth's magnetic field from the magnetic results of our measurements. Different results raise the basic question of whether the different strength of the Earth's magnetic field reconstructed from rock samples for different points in Earth's history is due to either unreliable recording of the signal in the rock samples or rather the strength of the Earth's magnetic field actually having fluctuated over a period of observation, possibly depending on the location.

What does a scientifically sound consideration of alternative theories look like?

A random example: When measurements on rock samples first provided evidence that the Earth's magnetic field had a different polarity several times in Earth's history than it does today, the magnetic *self*-reversal of the recording minerals in the rock was discussed instead. As an alternative solution to causal changes in the EMF[24]—somewhat as if some cameras occasionally, unnoticed, deliver mirror-inverted photos. Without sound and above all multidisciplinary research to scrutinize any alternative explanations, a thesis with misleading implications for the Earth system would probably still stand as the sole explanation about 100 years after these discussions.

No, researchers in the field of paleomagnetism do not necessarily have to work in all facets of (geo-)magnetic investigations or even possess special expertise. However, the scientific results of the respective disciplines in their entirety and mutual linkage form the foundation for ever further—refined or entirely new—insights in paleomagnetism. Depending on the emphasis, one might move more in the field of rock magnetism (How do rocks do that with magnetization?), geomagnetism (How is the Earth's magnetic field created?) or even archaeomagnetism, the recent history of the Earth's magnetic field, so to speak.

---

[23] This is the area of physics that deals with the behavior of nature at or below the size of atoms. Many laws of nature familiar from daily, macroscopic life—keyword *Sir Isaac Newton*—do not necessarily apply there.

[24] For example, Néel, L. (1951). L'inversion de l'aimantation permanente des roches. *Ann. Géophys.*, 7, 90–102.

To illustrate this with an example from everyday life: A successful medical therapy is based to a certain degree on research findings from genetics or evolutionary biology. The people involved in prescribing the therapy do not necessarily have to have been involved in the underlying evolutionary biological research. But they must incorporate the *insights* of these neighboring disciplines meaningfully into the therapy they are designing.

So one might classify paleomagnetism as a research direction in which the recording and deciphering of past states of the Earth's magnetic field is paramount. Such recordings of the Earth's magnetic field have two notable peculiarities.

First, there is hardly any other, *past* state of the Earth, whose geophysical parameters have been recorded over geological times and also worldwide, simultaneously and naturally, and are thus accessible to geoscientific research today.

Many geological parameters can be narrowed down by deduction. For example, the average surface temperature of the Earth in the geological past can be approximated from a wealth of clues. Useful indicators range from supposedly preferred temperature ranges of former life forms (flora and fauna) to astronomical factors such as the tilt of the Earth's axis (Chap. 2). However, there is no way to directly measure a temperature value archived from the geological past of our planet.

In paleomagnetism, researchers have the opportunity to devote themselves to the measurement of the so-called paleo*intensity* of the Earth's magnetic field, i.e., the analysis of a numerical value of a physical parameter of the Earth—many millions of years ago—from a rock sample. This is not a simple measurement, and more elaborate laboratory procedures, some math and physics, are necessary—but it is possible. Unlike measuring a temperature value in the Earth's past, as mentioned.

Second, both the intensity and the *direction* of the Earth's magnetic field are stored in rocks. In mathematics, the combination of strength (intensity, importance, size, etc.) and direction is referred to as a *vector*. A vector is symbolized by an arrow.

An arrow points in a certain direction. A prerequisite for naming a direction is always an underlying *reference system*. The direction of the Earth's magnetic field is usually given as an angle in degrees of declination and inclination, usually in reference to the known reference system of the Earth's longitude and latitude. The declination then tells us how much a magnetic direction deviates from the geographical longitudes (which, as we know, all point north; Fig. 3.11 and 2.7).

An important point: A (magnetic) direction, as it literally is out there, would not change or be rotated in any way if a different reference system were chosen. It would just be labeled differently. For example, we choose the walls of our (someone's) garden instead of the Earth's longitude and latitude as a reference system. The designation of the magnetic direction, i.e., the angles of declination and inclination, would now indeed be completely different, since, for example, the angle of declination is no longer represented as a deviation from the direction of the longitudes, but as a deviation from the orientation of our garden wall.

Another example: Even a residential building itself would remain unchanged if, for some reason, it was decided to change the house number. This could be because the new

3.4 Self-Reversal, Reference Systems and Arrows   83

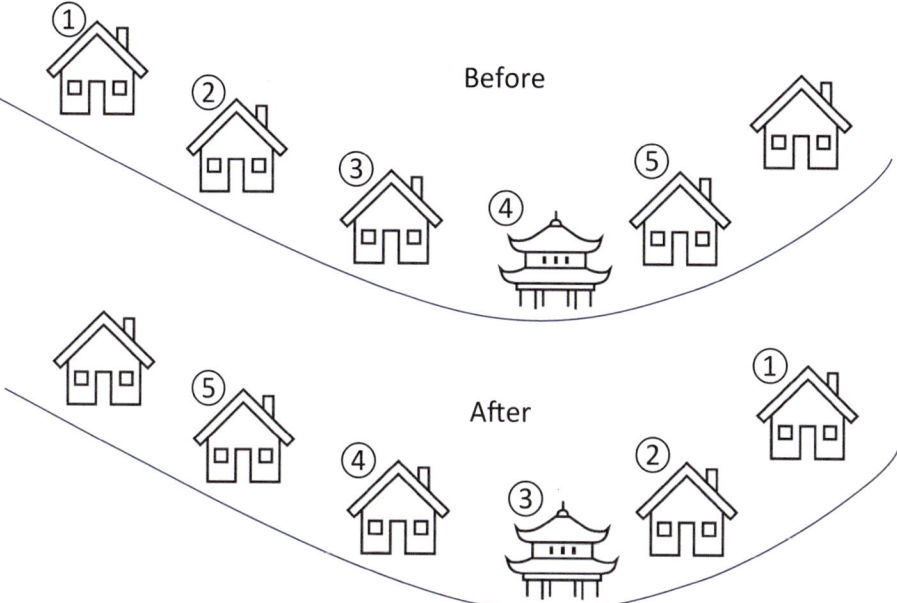

Fig. 3.12 The location of a house (before) does not change when a different principle for assigning house numbers is introduced (after)

numbering of the houses on this street is now supposed to start at the other end of the street and is therefore counted differently (Fig. 3.12).

In all cases, including the always somewhat limping analogy, the reference system, so to speak, rotates around the arrow (the house), not the other way around. The crucial point is that the renaming does not change the location or position relative to other houses (arrows, vectors).

The idea can be further developed. If I know the location of our garden wall in the geographical system of longitude and latitude, that is, I know that a wall runs from north to south and the head side from east to west, I can represent any (magnetic) direction both in relation to the garden wall and in relation to the cardinal directions. Even more importantly: From one representation, I can mutually derive the other specification. Mathematically, this would be a coordinate transformation.

In the reorganization of house numbers, we assumed a regularity in the new numbering—upper end of the street instead of lower end, while the number of houses does not change. So, a rule can certainly be found according to which the new house number can be calculated from the old house number. Old house number plus x, for example. In a coordinate transformation, we will usually not only consider one (up and down the street), but three coordinates (north-south, east-west, up-down or more generally an x-, y- and z-direction) to determine the direction of an arrow (Fig. 3.13). But that is basically the only difference.

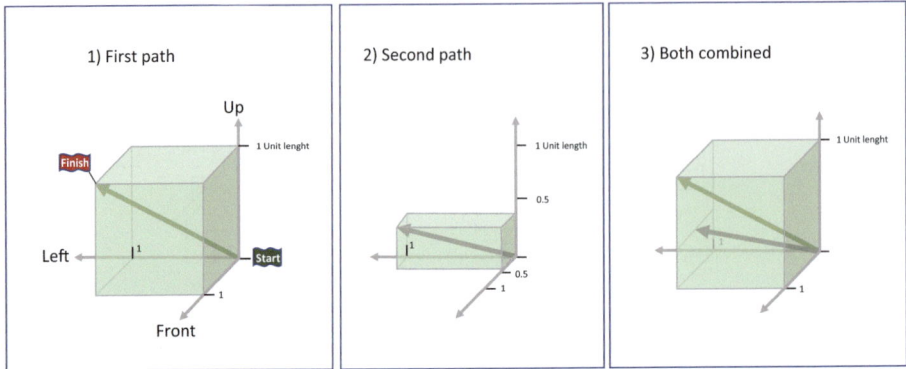

Fig. 3.13 Graphic illustration of the terms "vector" and "reference system". To get from the start to the destination in the left illustration, move one unit of length in any order along the three axes forward, left, and up. The three axes define the so-called coordinate system in which our path is represented. This direct path is shown by the arrow (vector)

In paleomagnetism, such coordinate transformations are frequently carried out—for good reason and often because it is simply practical, for example when taking rock samples (the garden walls) for magnetic measurements and the later representation of the magnetic result in relation to the system of longitude and latitude of the earth.

The direction of our arrow does not yet make a vector. An arrow also has a certain *length,* to characterize the strength with which something occurs in a certain direction (Fig. 3.14). An example from everyday life: If the wind blows in the same direction, let's say, from the front, it makes a significant difference for the bike tour whether a "gentle breeze" (low intensity, short arrow against the direction of travel) blows or we are cycling against a storm gust (high intensity, correspondingly longer arrow according to the difference in wind strength).

In relation to the Earth's magnetic field, the scientific term (see above) is not length or strength—the correct term in relation to the properties and effects of the Earth's magnetic field is intensity. We need to distinguish or at least briefly characterize some facts

Fig. 3.14 Illustration of the term "vector" using the example of tailwind, headwind, and wind strength. A combination of direction and strength can be represented by (directional) arrows, with differences in intensity symbolized by proportional arrow length. Left and middle: Same direction at different strength. Middle and right: Same strength at different direction. Vectors often encounter us in paleomagnetism to represent the direction and intensity of a magnetic field or the magnetization of a rock sample. Unlike the wind, we humans lack the senses for magnetic vectors (see also Fig. 1.2)

3.4 Self-Reversal, Reference Systems and Arrows

to prevent misunderstanding. Otherwise, we will not talk much about the intensity of the Earth's magnetic field in this book, only its directions:

- Not only does the direction of the Earth's magnetic field change in various ways, as indicated, but the intensity of the Earth's magnetic field also fluctuates measurably, both temporally and spatially. Simply put, the direction of the Earth's magnetic field can be the same, regardless of whether the field is stronger or weaker. Just like wind blowing stronger or weaker from the same direction. The reason for the fluctuations in the intensity of the Earth's magnetic field lies in the complexity of the underlying geodynamo processes, very similar to the explanation for the changing polarity. Recent measurements of the intensity of the Earth's magnetic field show, when viewed globally and similar to not time averaged directions of the field, notable variations and no particularly symmetrical pattern in their distribution. But even at the same location, the intensity can fluctuate by around 100%.
- In the geological past, too, the intensity of the Earth's magnetic field was not constant—apart from the direction measurements on rocks, this so-called paleointensity of the Earth's magnetic field can also be read from some rocks (and artifacts; more on this shortly). However, this requires special laboratory studies[25], because the paleointensity does not directly follow from the "length of the arrow" of our direction measurements. The "length of the arrow" depends very much on the amount and type as well as the same orientation of the magnetic particles in a rock. For example, a basalt will already give a stronger magnetic measurement because it contains many times more magnetite particles in a sample than, for example, in a limestone. However, both rocks would have recorded the same field direction at the same time of formation. Even if possibly with different accuracy.
- Lava can be an ideal rock for investigating paleointensity, as these rocks acquire their magnetization as TRM (thermoremanent magnetization; Sect. 3.2), that is, the orientation of the magnetic particles occurs during cooling along the magnetic field lines. The stronger the field, the more effective the alignment. If we want to imagine the magnetic directions again as arrows, the most effective alignment is the one in which all arrows point in the same direction (Fig. 3.15). A weak field does not manage to turn *all* arrows in the same direction even when a lava cools—the prevailing direction remains the same, but not all participate, if you will. An even weaker field will align even fewer arrows, and so on.

---

[25] For those who want to read up: A keyword search under *Thellier* experiments offers an introduction. In experiments to determine the paleointensity, rock samples or artifacts are heated and cooled in the laboratory and exposed to different magnetic field strengths. The acquired intensity of the stored magnetization is then compared with the original magnetization of the sample acquired in the Earth's magnetic field (a little calculation is involved) to infer the intensity of the field at that time.

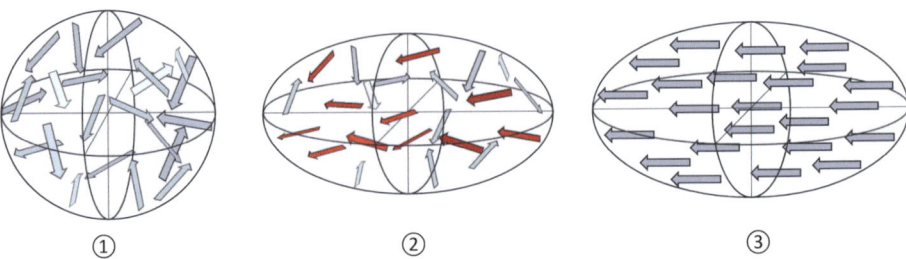

Fig. 3.15 Schematic representation for a better understanding of effective alignment to represent a direction. If in a group of arrows ① all point in different directions, there is no preferred direction. If a smaller group of arrows points in a similar direction ②, but the remaining arrows do not, a preferred direction emerges. A direction cannot be indicated more clearly than that all arrows point in the same direction ③. Even in a maximally *magnetized* state, all magnetic moments (arrows) are aligned. In some geological situations, in some types of rock, the direction of the Earth's magnetic field is rather captured in the manner of ② than in the most effective manner of ③. A magnetic preferred direction emerges as soon as several "arrows" point in the same direction. All other arrows are still randomly oriented

- Since a few thousand years ago, when people began to fire clay goods, pottery or the walls of the kilns (also built of stone) of these artifacts can also be used for measurements of paleointensity. Ultimately, the clay—clay naturally contains magnetic particles like magnetite—is heated above the Curie temperature of magnetite (580 °C), cools in the Earth's magnetic field and thereby stores the direction and intensity of the prevailing Earth's magnetic field according to the principle of TRM (Sect. 3.2). In the case of the kiln walls, the acquired direction is overwritten again and again until the kiln is used for the last time and this direction then remains preserved in the walls of the kiln.[26]
- Similar to the creation of the GPTS (see under "The geomagnetic polarity time scale" in Sect. 2.2), a precise knowledge of the course of the paleointensity fluctuations of the Earth's magnetic field can be used to classify geological events in Earth's history.

In Fig. 1.2 the vector property of magnetization is illustrated in a different way; however, unlike the wind analogy in Fig. 3.14, we usually do not feel magnetism, so we now have a usable experience from everyday life to approach vectors in geomagnetism, but still not the whole picture. With our new understanding of the preference of paleomagnetists to draw arrows, the further explanations, especially in Chap. 4, are easier to understand.

---

[26] For example, Clarke, A.J., Tarling, D.H., and Noël, M. (1988). Developments in archaeomagnetic dating in Britain. *Journal of Archaeological Science*, Vol. 15, Issue 5, 645–667.

3.5 Magnetic Susceptibility of Rocks and Minerals

After the directional remarks in the previous section, we should at this point take a closer look at the initially quoted magnetic susceptibility of rocks and minerals, or rather, put it into an illuminating context, mainly in preparation for Chap. 5. As stated, the blanket magnetic susceptibility is a material/mineral-specific characteristic and has per se no necessity for a geomagnetic field or its direction-related concerns. Without susceptibility, however, the magnetism of rocks is of course unthinkable.

The property of magnetic susceptibility of rocks changes, sensitively, measurably at least in the centimeter range, with every change in the composition of the rock. To the eye, such a change in a rock may lie in a slight change in color or, for example, in a change in the dominant grain size of a rock. Often changes in the composition of a rock are not recognizable to the naked eye. Anyone who has had the opportunity to look at hundreds of meters thick, monotonously gray clays from a deep-sea drilling will agree here. Such seemingly monotonous deposits are by no means monotonous upon closer inspection. Rather, even in macroscopically monotonous rocks, the finest changes in the composition of the rock can be made visible with suitable methods—magnetic susceptibility is one of the possibilities.

For classification: Typical clastic[27] sedimentary rocks are formed in a kind of natural recycling process of existing rocks. Such sedimentary rocks are solidified at the earth's surface from weathered, then eroded minerals and fragments into a rock again (Fig. 3.16)—after these components have been transported in various ways over different periods of time before they were deposited again.

Where the sedimented material comes from and how it is delivered—there are countless possibilities. Often, material from various sources (i.e., eroded rocks) is ultimately collected during deposition. In a geological scenario, such as the formation of a sediment from river deposits, the change in the composition of the deposited rock could reflect the slow meandering of the river over thousands of years. Or a change in the source area of the river and thus the sediment load carried along. Sometimes several rivers from different geographical directions, i.e., erosion areas, flow into the same lake. When depositing the transported fragments, a change in the relative amount and type of delivered clay minerals is enough to change the rock properties in the millimeter range of this new layer. If a sedimentary rock, for example, consisted of local lake deposits, even a seasonal change of washed-in organic material[28] before solidification into a rock could

---

[27] No, not *classical* sediments, but those sediments which mainly consist of *clastic*, i.e., mechanically eroded components of earlier rocks.

[28] With the transported fragments of earlier rocks, organic material in the form of plant remains, animal remains, and microorganisms will always be delivered. When the material is slowly solidified into a sedimentary rock, organic material is certainly less durable than, for example, a quartz grain. As we will read in Sect. 5.2, the degradation of organic material in developing rocks in turn triggers processes that also and especially affect the magnetic minerals.

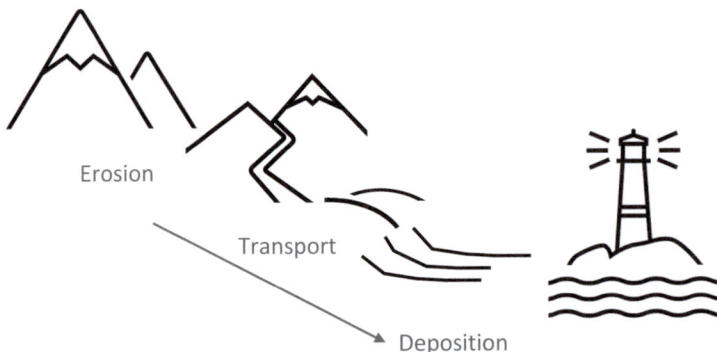

Fig. 3.16 Clastic sedimentary rocks are formed by weathering and erosion from all types of existing rocks, for example other sedimentary rocks, but also magmatic granites, basalts or from metamorphic rocks altered by high pressures and temperatures of a mountain building process. Weathering separates the individual mineral grains, erosion sets the grains on their way following gravity. Depending on the length and type of transport path, the mineral grains are further crushed, rounded and possibly completely dissolved depending on their resistance. The length of the transport path can vary between a few meters and several hundred kilometers depending on the geological and topographical situation. As a general rule, we expect that a longer path will also lead to a stronger change (crushing, rounding, "dissolution" and thus change in the composition of the components) of the transported material. Quartz is one of the robust minerals of the earth's crust and therefore usually "makes it" as the only component to the deposition at the (sand) beach. Understandably, eroded components are transported from a topographically higher location to a lower one following gravity. It does not always have to be, as symbolized in the figure, the maximum path from the high altitudes of a mountain to the sea. Material can also be transported from a surrounding hilly landscape into a lake or in the direction of the wind through the desert. Often a depositional environment has several catchment areas with long-term changing influence on the composition of the resulting sediment. Imagine sediment-carrying rivers that converge into a main river at different locations and whose material is then transported together from now on. As mentioned at the beginning, our earth's crust is subject to dynamic change over geological time periods, and so the newly deposited sediments are solidified and can be piled up into a mountain again in a plate tectonic process, starting a new sediment cycle.

cause a corresponding change in the rock composition. Any change in the composition of the rock also changes the value of the magnetic susceptibility.

Sometimes a rock is secondarily altered, long after the original solidification, for example, by waters circulating in the underground and reacting in manifold ways with the rock components (mineral grains). We will elaborate such a scenario elsewhere; here we want to note that even in this scenario a, at least locally similar and thus magnetically detectable, change of the rock occurs. Components of the rock are altered or, for example, washed in as micrometer-sized, new mineral components into existing cavities (pore space) of the rock. Every component of a sedimentary rock, whether mineral grain, binding cement, organic parts from shells or plant remains, even water bound or free in the pore spaces, contributes to the measured value of the magnetic susceptibility of a rock sample. The respective contribution to susceptibility can be strongly positive, rather small, or in some cases even negative (Fig. 3.17 and 3.18).

3.5 Magnetic Susceptibility of Rocks and Minerals

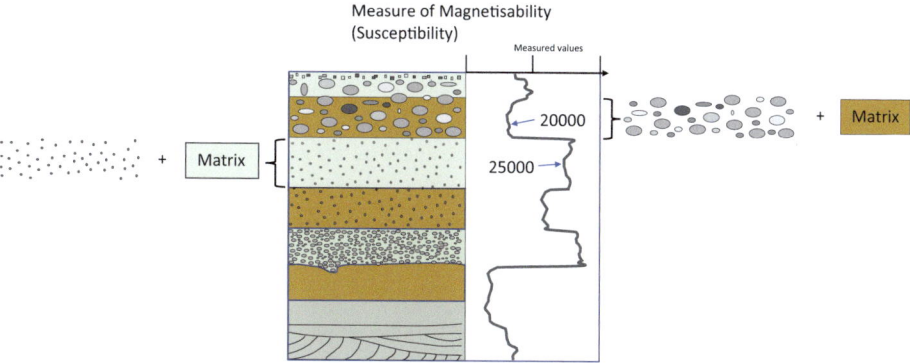

Fig. 3.17 A closer look at the fictitious curve of susceptibility measurements over a series of sediment layers. Depending on the composition of the components of a layer, the strength of the magnetic susceptibility at a certain measuring point changes. Two measuring points are highlighted in the figure, and we have assigned them values of 20,000 and 25,000 (in the same, for simplicity's sake deliberately ignored, correct physical unit of magnetic susceptibility). For simplicity, we further assume that all grains of the same mineral (different gray shading of the grains) in a respective layer have exactly the same size and shape (which is not impossible in a geological scenario for the deposition of this layer, but admittedly extremely unlikely). As matrix we refer to everything that is not present as macroscopically visible grains of this layer, for example the (carbonate or silicate) binder cementing the grains. Every component of the rock in the measuring range contributes to the measured value of susceptibility—by a value characteristic for this component and the amount of the respective component (see also Fig. 5.20)

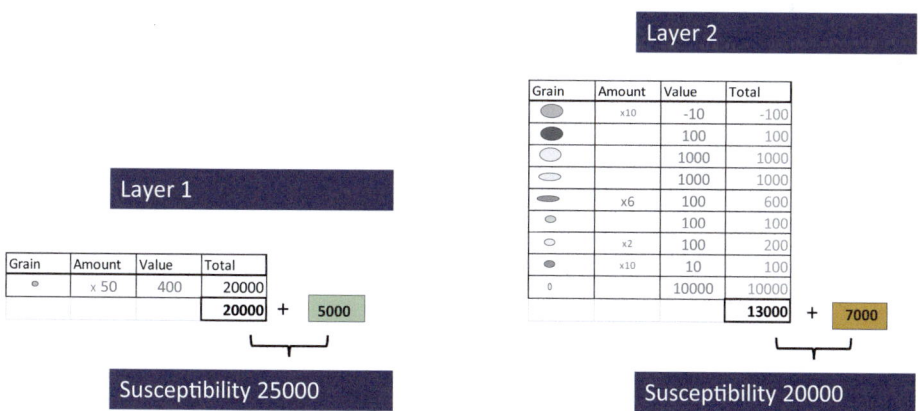

Fig. 3.18 For understanding the (fictitious) susceptibility values in Fig. 3.17. The bulk susceptibility value results from the relative contributions of the respective components of a rock (right/left table for the respective layers in Fig. 3.17). Shown are fictitious values to illustrate orders of magnitude

Quartz has different values than feldspar or mica; as a basic rule, the respective contribution to susceptibility depends on the type of material and its quantity. With equal amounts[29] of quartz and mica in a rock sample, the respective contributions to the susceptibility value of the measured rock sample are significantly different due to the different magnetic properties. Figure 3.18 explains these relationships in more detail. Neither quartz nor mica have magnetic properties in the classical sense, as we imagine a magnet—not even (attention!) in the sense of the remanent magnetic minerals from Sect. 3.1, which can record the Earth's magnetic field. Quartz, like most other minerals in a rock, does not contribute at all to the magnetic analysis of the Earth's magnetic field in the geological past—all these minerals, however, react magnetically differently than other types of minerals, *as long as* they are exposed to a magnetic field. This happens, for example, through a suitable measuring device when measuring magnetic susceptibility. Water also "reacts" in this sense to a magnetic field, without being able to record it. However, the "reaction" is as low as quartz in the world of magnetic measurements. A sensitive measuring device is necessary.

We want to better understand these "temporary magnetic reactions of matter" with the example of magnetic applications in Sect. 5.2 to Bamiyan. Many other similar applications are conceivable and are indeed used in the geosciences. Once we have explained the principle of the application, it will be easy to understand such other applications, for example the correlation of drill cores in the deep sea or even the monitoring of air pollution in a city. In each of these applications, the magnetic measurement value correlates with the complexly composed content of a "black box". They characterize this box, without us necessarily having to know the content, or more precisely, every single component.

Our (arbitrarily assumed) dimensionless measurement value of 20,000[30] results from the sum of the individual contributions of all grains captured in the measurement range plus the matrix. Simply put, the matrix refers to particles that are so small and numerous that we no longer refer to them as individual grains, but rather as a kind of filler between the visible grains. We have symbolized various mineral grains of different sizes in Fig. 3.18 under the category "Grain"[31]. The "Value" category assigns a strength to each "grain" that characterizes the respective mineral type and thus the magnetizability of these grains. This illustrates in addition to Fig. 3.18 for explaining magnetic properties the orders of magnitude of magnetizability (magnetic susceptibility):

---

[29] Scientifically, it is more correct to speak of volume instead of quantity here.

[30] This value and the following values of susceptibility are completely arbitrarily assumed and illustrative. In reality, rock magnetic measurements are more likely to encounter very small numbers on the order of 10^{-6} (one millionth), but this does not change the principle of the explained.

[31] For simplicity, we use terms like "gravel", "grain" or "particle" in this book not in their scientifically strict definition, but to refer to the components of a sedimentary rock. "Gravel" and "grain" in everyday language rather evoke the association with coarser components, "particle" with finer components.

3.5 Magnetic Susceptibility of Rocks and Minerals

- Diamagnetic materials like quartz or water (weakly negative), here exemplarily as value: -10
- Paramagnetic materials like olivine or feldspar (positive), here exemplarily as values: 100, 1000
- Ferromagnetic (s.l.) materials like magnetite or some ores (strongly positive), here exemplarily as value: 10,000

Since the contribution of each gravel in the measurement range also depends on the respective amount of this type of gravel captured, the contribution to the measured susceptibility in the "Total" category results as a product of the factors "Amount" and "Value". The first listed gravel may be quartz, whose ten grains captured in the measurement contribute a diamagnetic contribution (strength -10) of $10 \times -10 = -100$ in the "Total" category. The last of the listed gravels may be a single, tiny grain of magnetite. Due to the strong ferromagnetic susceptibility, this grain captured in the measurement contributes 10,000 in the "Total" category to the measurement value. All contributions of the gravels added together give the value 13,000. Also captured in the measurement is the (brown) matrix that holds the grains together. Here too, the contribution results in the same way as a product of quantity (volume) and specific factor of the material type. We have assumed the value 7000 for our example. The sum of the gravels (13,000) plus matrix (7000) gives our assumed value of 20,000, as initially stated.

On the left in Fig. 3.18, it is shown how the higher value of 25,000 could result at another measurement point, in another layer. The matrix may consist of a slightly different material than the brown matrix in our previous example. We assume the contribution as 5000. Otherwise, only one type of gravel is found in this layer, of which 50 (paramagnetic) grains with a characteristic magnetizability of 400 each. 50×400 results in 20,000 plus matrix the measurement value at this point of 25,000.

So, this is the principle by which a susceptibility value is measured. As mentioned, the assumed values and quantities are purely illustrative, to give a sense of magnitudes and ratios. In this way, it becomes understandable that a small amount of strong ferromagnetic components can make a dominant contribution to the measurement result due to their high magnetizability, for example, a magnetite grain in the sole vicinity of quartz grains in a sandstone. The examples hopefully provide enough basis to visualize the result of the many possible combinations in rocks. In particular, we should now understand how even the slightest changes in the composition of a rock can change the magnetic susceptibility. In our simplified example, it is enough to exchange one or two grains for others to change the total value of the susceptibility at this measurement point, albeit slightly. This principle is the basis for assuming magnetic susceptibility as a proxy for different rock layers (Sect. 5.2).

In summary, one could say in a transferred sense and rather banally: rye flour instead of wheat flour in the cake dough changes the taste of the baked goods as a result. More scientifically, the magnetic susceptibility of a rock sample results from the sum of the contributions of the individual components, i.e., from the mineral grains plus possible secondary components, such as pore water or plant and animal fossils in this sample.

In the context of this book, the following assertions should be noted:

1. Sedimentary rocks reflect in their composition the finest changes in environmental conditions at the time of sediment deposition.
2. Such changes in the composition of a rock can be made visible by magnetic measurements, for example, the magnetic susceptibility.
3. The deposition of sediments ideally reflects at least locally, if not a regionally similar signature of changing environmental conditions.[32]

---

[32] To illustrate these terms, one might think, for example, of the local boundaries of a lake compared to the more regional deposits along a long stretch of coastline.

Paleomagnetism—Applications 4

There are many ways to approaching the term paleomagnetism—classically etymological, for example, with reference to the meaning of the affix "paleo" from the (ancient) Greek *palaios* (παλαιός), meaning "old", "former", "erstwhile". We know the use of this affix from other, often scientific terms such as "paleontology" (the study of ancient beings, i.e., fossils, in the sense of deciphering past life forms and their environment), "Paleozoic" (ancient earth in the sense of the oldest time period in the development of living beings) or "Paleolithic" (Old Stone Age, in the sense of the oldest time period in the use of stone tools in human history since the Pleistocene).

So we understand that paleomagnetism refers to a branch of science that combines "old" and "magnet". Hmm. Even over millions of years, magnetism, as we physically understand magnetism (Chap. 3), has probably not changed, has evolved or even become extinct in the course of evolution. As has happened to not a few life forms in the course of Earth's history. A primarily temporal use of *palaios* does not sufficiently explain the term "paleomagnetism".

As discussed in Chap. 2, key characteristics of the Earth's magnetic field have continuously changed over the billions of years of Earth's history. The magnetic north pole has repeatedly become the magnetic south pole and vice versa. Current measurements show that in addition to direction, the strength (more scientifically correct: intensity) of the Earth's magnetic field is subject to considerable fluctuations.[1]

It is therefore likely that such fluctuations were also relevant in the geological past,[2] especially when considering the enormous impacts that a pole reversal would have on the

---

[1] International Geomagnetic Reference Field (IGRF), e.g., P. Alken, et al., 2021, International Geomagnetic Reference Field: the thirteenth generation, Earth, Planets and Space, Vol. 73, Article 49.

[2] A frequently cited credo in the geosciences states that processes observed today have always been the same—"always" is understood in the sense of "geological time periods", unless we find a reason to reconsider and revise such an assumption.

© The Author(s), under exclusive license to Springer-Verlag GmbH, DE, part of Springer Nature 2025
M. Urbat, *Magnetism of Rocks*, https://doi.org/10.1007/978-3-662-70428-8_4

intensity of the Earth's magnetic field, because there is no bar magnet in the Earth's interior that is occasionally turned around.

At first approximation, paleomagnetism deals with the reconstruction of past states of the Earth's magnetic field over the course of Earth's history. This would probably be less interesting outside of pure geoscientific research if it were not possible to draw further conclusions from these records about the development of the Earth system. Accordingly, we want to get to know some typical applications of paleomagnetic research that serve this overarching goal.

All geological applications in this chapter have in common that they refer to the known cardinal directions in the sense of north, south, east, west as well as "up" and "down". Directions are indeed often of decisive importance in the geosciences: The practical use of geoscientific research is permeated with questions about the flow direction of long-gone rivers to the permeability of a network of cracks in a hard rock. Such information is crucial today, for example, for the presence or absence of groundwater or for the presence or absence of raw materials. In this chapter, we will therefore address, among other things, the following questions:

- What does the Earth's magnetic field (EMF) have in common with fossils?
- Why are arrows so popular in paleomagnetism?
- How do you work with the barcode of the Earth's magnetic field?
- How do you read magnetic signposts (small and large)?

Less colloquially, we will try on the following pages to fill the term "paleomagnetism" with content in a way that allows us, in reference to the more physical insights in Chap. 3, to gain an insight into the geological and practical background of this branch of the natural sciences.

Rocks record the state of the Earth's magnetic field (EMF) at the time of their formation. How this is possible for rocks, we have discussed in more detail in Chap. 3. But if properties of the Earth's magnetic field such as direction and intensity have repeatedly changed in the geological past, then what was magnetically recorded in rocks has also changed. This is the foundation of the geomagnetic polarity timescale.

The EMF has a three-dimensional shape (Fig. 4.1 and 2.6), and part of this "shape" is recorded as a likewise three-dimensional, albeit "virtual object" in rocks. Throughout the book, we have compared this object, among other things, to a pencil (Fig. 3.4). Here at this point, we realize that the entire Earth's crust, including the (rocky) ground on which we stand and the mountains we climb, is a gigantic archive of the states of the Earth's magnetic field over the past millions of years. For paleomagnetism, a rock sample is thus a kind of magnetic "fossil".

Just like the "classic" fossils from the plant and animal world in paleontology, the best magnetic fossils are always those whose original information about the past Earth's

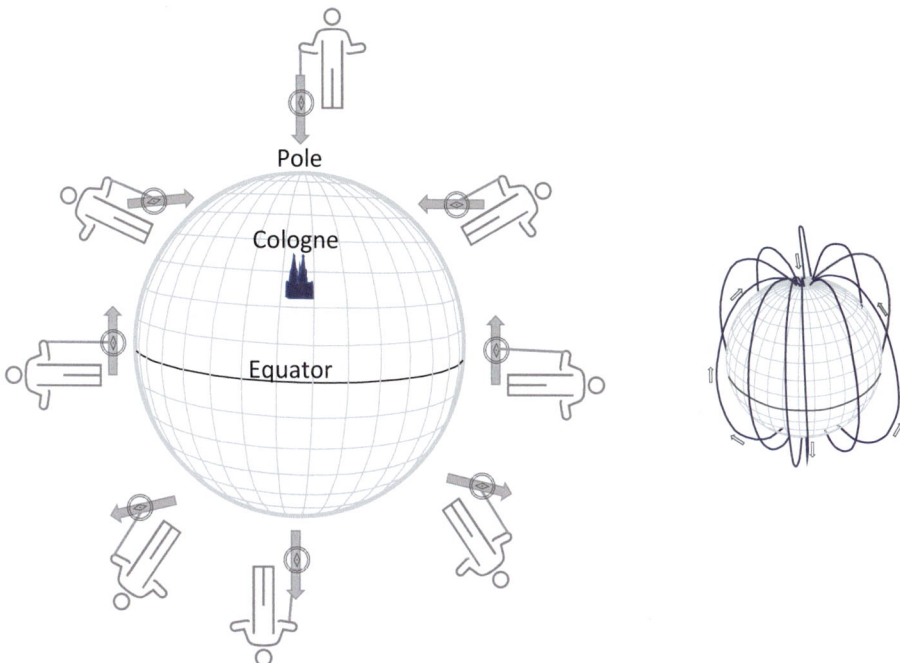

Fig. 4.1 Under certain conditions, there is a direct correlation between geographical latitude and magnetic inclination. This dependency becomes understandable when compared with the three-dimensional representation of some field lines of the Earth's magnetic field (small image on the right; see also Fig. 2.7): steeper inclination angles are correlated with increasingly higher latitudes, while the angles near the equator are almost flat, i.e., the field lines lie parallel to the Earth's surface (see also Fig. 4.2, 2.7 and 3.11)

magnetic field (their "shape") has been preserved as well as possible and has not been subsequently altered, deformed, or even erased.[3]

Very similar to the reconstruction of the organism and its environment from a fossil petrification, we will also have to pursue various ways of reconstruction in the magnetic sense through laboratory experiments, analysis, and deduction and induction (what would language be without wordplay) in order to be able to read Earth's history from the magnetism of rocks. In particular, we refer to Chap. 2 to explore how the repeated polarity reversals represent the timescale, as used in one of the most prominent applications of paleomagnetism: magnetostratigraphy.

Let's return to further direction-related questions in geology and mention a few more examples: In which way was the pressure directed in plate collisions, or, in other words,

---

[3] If not already mentioned, the Earth is a dynamic entity, whose rocks of the Earth's crust are little permanent when viewed over millions of years.

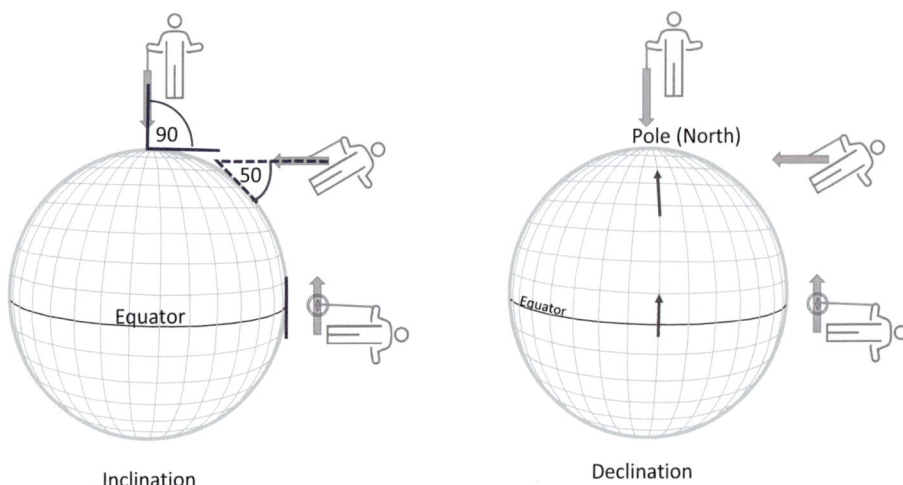

Fig. 4.2 In the case of a time-averaged EMF and the acceptance of the GAD[4], the magnetic inclination at the pole is 90° and decreases with lower latitudes until it is 0° at the equator. The magnetic vector lies parallel to the Earth's surface there. The declination (right) is 0° at every point under the assumption of the GAD. In other words, the field lines lie parallel to the Earth's longitudes

how did two Earth plates move relative to each other? In which orientation do corresponding fracture planes in the rock run?[5] In which direction do fluids flow through rock layers in the subsurface (Chap. 5)?[6]

The list could be continued for a long time. In addition to infrastructural applications, such as in the construction of a tunnel or a dam (stability of the rock), directional characteristics of rocks are of enormous importance in questions from earthquake analysis to the exploration of raw materials.

Such characteristics can certainly be measured in various ways in the existing rock, in the subsurface. But as soon as a drill core is recovered from the depth of the Earth's crust

---

[4] Geocentric axial dipole hypothesis, see Chap. 2.

[5] The formation and orientation of fractures in connection with tectonic deformation follow certain laws. It is important for the imagination to realize that a rock layer breaks in response to a load. In the Earth's crust, for example, due to tensile stress when a continent breaks, or due to pressure, for example, due to overburden or the collision of Earth plates. The direction and orientation of the resulting fracture planes are linked to the direction of the force applied. In other words, it is predictable in which direction a fracture in the rock will form if it is known in which direction the force is acting. Or vice versa. As a rule, not only one fracture plane is created in rock layers under load, but a whole set of equally oriented planes. Possibly in several, but still predictable directions. Also note that what we might see as cracks in a rock wall, when viewed three-dimensionally, is a fracture **plane**.

[6] See Chap. 5 for this.

or a sample is generally taken from a rock wall, the reference to the cardinal directions is lost in many cases. In which original orientation does the fracture surface in my rock sample belong to the further rock association in the Earth's crust? How can I therefore relate the information from the rock sample in my hand to the area of the Earth's crust from which the sample originates? All this is necessary information to be able to answer some of the direction-related questions listed above.

In another useful application, the magnetic signal, due to its representation of the direction-related Earth's magnetic field, also allows for a subsequent (re-)orientation of rock samples to understand the spatial position in which this sample was located before being removed from the rock association in the Earth's crust.

As we discussed in Chap. 2, we like to think of the Earth's magnetic field as a global structure, particularly symmetrical to the Earth's longitude and latitude (Fig. 2.7). Especially based on this geocentric axial dipole hypothesis (GAD; Chap. 2), the magnetization direction of the EMF recorded in rocks plays the mentioned role in many paleomagnetic applications.

In the appropriate case, the magnetization stored in a rock reveals the latitude on Earth where the rock was formed at the time of magnetization acquisition, as the latitude corresponds to a specific inclination angle of the field lines (Fig. 4.1 and 4.2). If the part of the Earth's crust where the relevant rock was formed has drifted from its position since the storage of the magnetic direction—keyword plate tectonics -, we directly understand a possible application: Bring the crust part back to its original position by searching for the latitude with the appropriate inclination angle of the magnetic field lines. But more on that later. Especially in practical implementation, some other things must then be considered, analyzed, and implemented.

There are a number of other ways to use the directions of the Earth's magnetic field stored in rocks from the geological past.

Geoscientists are very interested in drill cores. Certainly, there are some high-resolution, especially seismic methods, to "pictorially" represent the geological layers of the deeper Earth's crust. Even three-dimensionally and in very high resolution. However, being able to hold and examine an actual rock sample from the depth of the Earth's crust in hand to verify (sometimes even understand) the "picture" is irreplaceable.

The ocean floor, for example, has been "cored" many hundreds of times to date—thus not nearly often enough for all outstanding geoscientific questions—up to hundreds of meters deep. Ocean-going international research ships like the *JOIDES Resolution* of the IODP (*International Ocean Discovery Program*) for example, have been contributing significantly to our modern understanding of the Earth for decades. The mentioned ship is over 140 m long, with a 62 m high drilling tower and packed with sensitive geoscientific laboratories. Including per expedition invited scientists, technicians, and crew, the ship is occupied with about 160 people on its approximately two-month research trips.

Funded and coordinated by a multitude of academic institutions and national research organizations[7], many months of planning go into each expedition. Sophisticated engineering methods are ready to bring drill cores from hundreds to thousands of meters of water depth plus hundreds to thousands of meters of drilling into the ocean floor to the Earth's surface. However, a complete orientation of the drill core often has to be omitted for reasons of practicability or simply cost-benefit reasons.[8] We will learn in the following that the stored Earth's magnetic field in rocks not only provides an age signature, but also a kind of built-in direction indication in rocks.

4.1 Magnetostratigraphy

Why Dating is Important

Stratigraphy is an important aspect of geology, as this term encompasses all investigations dealing with the temporal and spatial sequence of rock layers in nature—very similar to the use of the term in art history to describe (gr.: *graphein*) the sequence of individual color *layers* (lat.: *strata*). In geology, there are a number of further specifications to better describe the focus of layer characterisation, for example *chrono*stratigraphy with a temporal perspective or the closely related *bio*stratigraphy, which in the broadest sense uses the (also formerly) organic components of our environment for the subdivision of rock layers.

In *magneto*stratigraphy, rocks are related to each other in time and possibly space based on the past states of the Earth's magnetic field recorded by the rocks. All dating methods, including magnetostratigraphy, serve to place the geological history of the Earth as precisely as possible in time. The previous considerations on the geological time scale in Chap. 2 make clear that the temporal classification of Earth's history in all its nuances is an extremely extensive and complex scientific challenge. The process is, so to speak, in no way set in stone. It requires the work of various research branches and has been ongoing for decades to advance the "Geological Time Scale" project. Specifically for magnetostratigraphy, we ideally need the complete knowledge of all pole reversals in geological history. A practical example follows.

---

[7] In total, in 2024, the American *National Science Foundation* (NSF), the Japanese MEXT, and the European ECORD (including the *German Research Foundation*, DFG) and over 20 other nations are involved.

[8] During the drilling/coring, a drill core rotates in the hollow drill bit around its (vertical) axis. Usually, "top and bottom" of the drill core can be reconstructed. However, a crucial component of the orientation often remains unknown. Drill cores often break during drilling, and individual core pieces rotate "uncontrolled" around their longitudinal axis. Now, with all the effort to get such a rock sample from the depth of the Earth, it would be more than regrettable not to know in which complete orientation this sample was stuck in the depth in the rock association (see list of direction-related geological questions).

4.1 Magnetostratigraphy

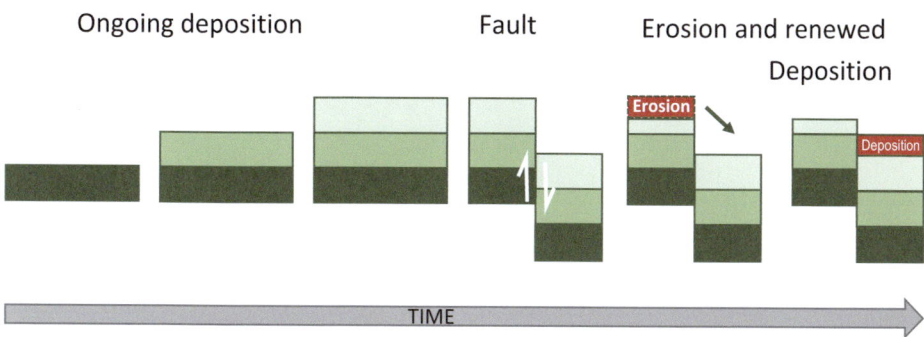

Fig. 4.3 Timeline of the fictional geological scenario discussed in the text. 1) successive deposition of three sediment layers, 2) subsequent tectonic disturbance, 3) partial erosion and re-deposition of sediments. In Fig. 4.4, 4.5, 4.6, 4.7, 4.8, 4.9 and 4.10, the processes are further illustrated

Discussion of a virtual, geological investigation

To better understand the challenges of (magnetic) age dating of rock layers, let's consider a simplified geological scenario (Fig. 4.3, 4.4, 4.5, 4.6 and 4.7) from the context of sedimentary rocks (sediments). As a reminder: All rocks of the Earth's surface down to depths of several tens of kilometers are old by human standards, often millions of years old. However, these rocks were not "always there", but they were formed, in the case of sediments, under specific environmental conditions in a certain deposition space. These clastic sedimentary rocks have often been deposited over long periods of time as initially loose fragments, grains and other particles and, for example, successively wedged into each other due to increasing overburden or solidified by circulating waters in the interstices and precipitated binding agent.[9]

In the case of an undisturbed first deposition of sediments, a simple geological basic rule (cf. Nicolaus Steno[10]) states that older rocks lie beneath the younger ones. For example, anyone who has a compost heap in the garden intuitively understands this rule: The next layer follows with the next load.[11]

---

[9] Alternatively, one might imagine the formation of a magmatic rock, where molten magma from the Earth's interior cools and solidifies in areas closer to the Earth's surface. As we discuss in Chap. 3, a specific time of formation can be designated for each rock.

[10] About 350 years ago, the universal scholar born in Copenhagen formulated some basic laws of modern stratigraphy—on the deposition of sediments in geology—after scientific studies in Italy and published them in 1669 in a dissertation De solido intra solidum naturaliter contento.

[11] However, there are countless geological scenarios in which this simple age rule is questioned or at least complicated. To explain, another analogy should be used. There are supposed to be desks of creative individuals where new books, articles, and magazines continuously accumulate. We briefly ignore the existence of electronic editions and the more modern accumulation on a storage medium. The current articles—recognizable by the publication date—come to lie on top of the stack. Depending on the speed and quantity of incoming new arrivals, the probable position of articles deposited in the past could be estimated. This works until the point when a book is pulled out from the bottom, the stack tilts, and the tower with its components, i.e., articles and magazines, spreads over the surrounding stacks. In geological terms, this is referred to as tectonics, i.e.,

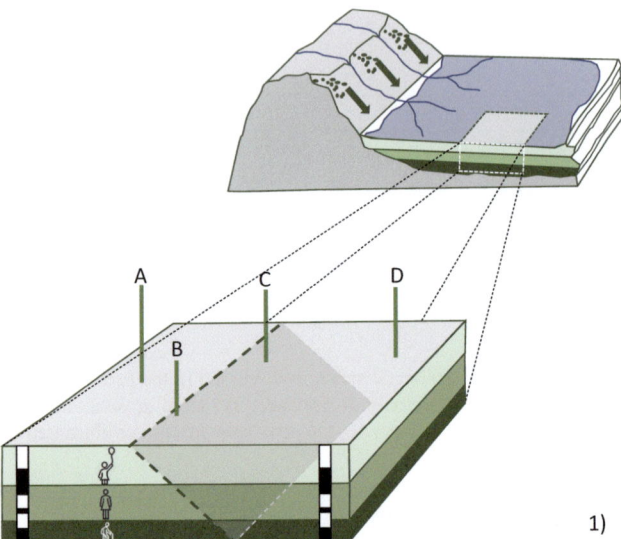

Fig. 4.4 Possible geological scenario leading to the deposition of the three layers Senior, Person, and Child. The naming of the layers is intended to simply illustrate the relative age of the layers to each other. In the geological situation shown, sediments are deposited in a "basin" (top). During deposition, the layers store the prevailing direction of the earth's magnetic field (black-white pattern, bottom; see also Fig. 4.10). Four fictional drill sites (A to D) are marked in their position. The dashed area around points B and C marks where a tectonic fault will displace the layers (and thus the magnetic records) against each other in the future (Fig. 4.5)

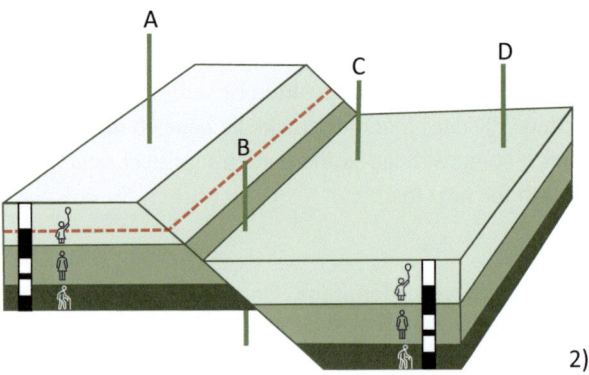

Fig. 4.5 At a displacement, the three layers were largely offset against each other. The drill sites A to D would have produced similar cores in the scenario in Fig. 4.4. This is now obviously no longer the case, especially not with the drillings B and C. In the situation shown, these drillings would penetrate the fault surface. The differences, especially on the respective cored magnetic profile, are summarized in Fig. 4.7 and discussed in the text

---

large-scale and small-scale movements in the earth's crust. These are extensions, compressions, and tensions, which are caused by the constant relative movement of individual parts of the earth's crust. Of course, it might be more effective in the long run to tidy up the desk at the end of each day. However, there is no such tidying up in geological scenarios.

4.1 Magnetostratigraphy

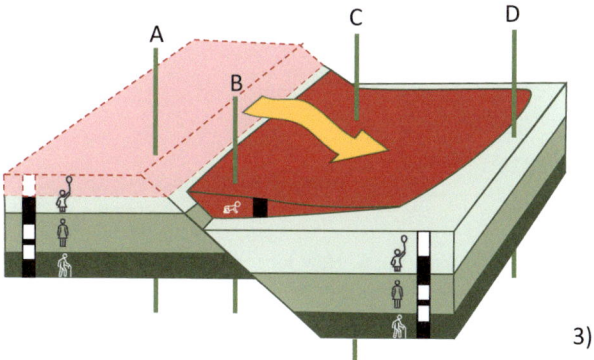

Fig. 4.6 Erosion of part of the toddler layer on the left, higher-lying block due to tectonic offset. The eroded material is redeposited on the right, lower-lying block following gravity and relief. Note the impact of this geological process on the stored magnetic signal in the layers. In particular, the erosion on the left block will destroy the original magnetic signal. During the redeposition, a new signal (the then prevailing situation of the Earth's magnetic field) is also recorded (see also Fig. 4.7)

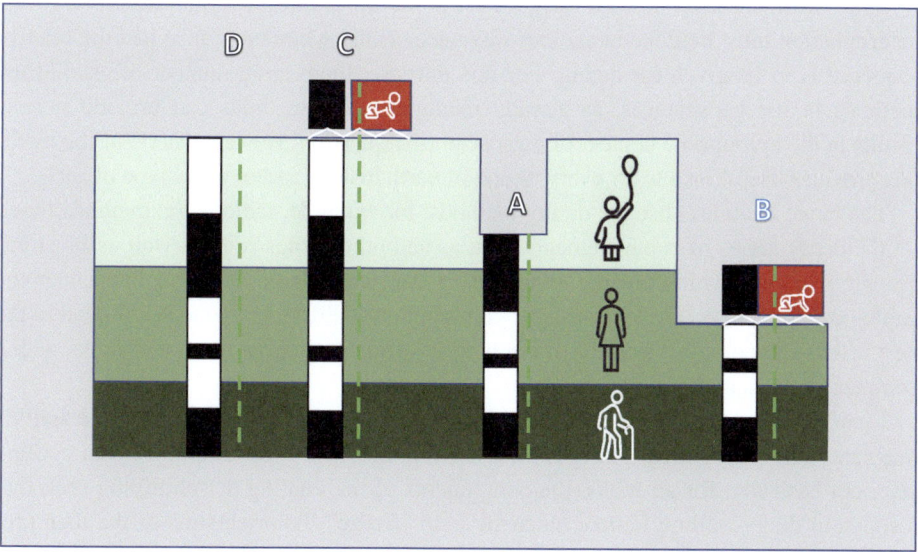

Fig. 4.7 Schematic representation of what the four boreholes A to D would yield in the geological situation in Fig. 4.6. Borehole A is incomplete, the upper part of the child layer is missing. Borehole B yields the smallest proportion of the original sequence of the three layers. Instead, in the upper area, the baby layer deposited with a temporal gap (indicated by the white zigzag line) is captured. Borehole C captures the entire original sequence of layers as well as the discordantly deposited baby layer. Borehole D yields a profile of the layers, as it would have been yielded immediately after their deposition (Fig. 4.4) (see also Fig. 4.8)

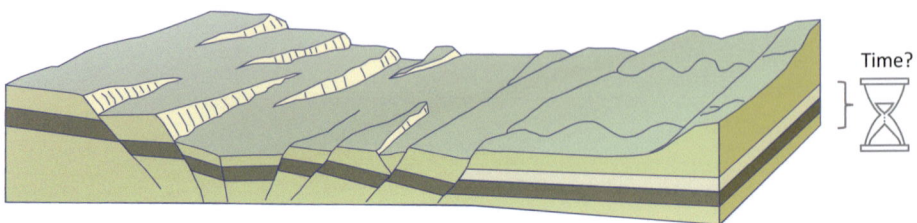

Fig. 4.8 Rock layers form at different rates. How long it took to deliver "the material" to build up a certain thickness of a rock varies considerably from sediments to magmatic rocks. But even among the sedimentary and magmatic rocks, the range of possible durations is enormous. How much time is therefore in a—for example 1 m thick—rock sequence, must be investigated in each individual case. The text contains examples of orders of magnitude of the rates at which sedimentary rocks form. On the right side of the sketch, the layers are horizontal, i.e., in the orientation in which these layers originally formed. On the left, a subsequent tectonic displacement of the layers is indicated; with a little geological practice, one recognizes a stretching of the Earth's crust in this area. Along individual faults, individual blocks have slipped against each other and tilted slightly. The tectonic displacement occurred long after the original deposition of the layers and in no way changes the "contained time" in the exemplarily sketched, dark layer (see also Fig. 4.9)

Since sediment layers, unlike magazines, do not have a printed publication date, age determination must be done in another way—especially when more than just the relative sequence is to be given for dating. For this purpose, there are a number of established methods in the geosciences, as already mentioned. All methods can provide precise results in the appropriate context of a geoscientific question. However, none of the methods provides useful results for every period of earth history and/or every type of rock.

The range includes absolute dating methods, for example, radiometric methods based on the decay series of some elements such as carbon or potassium, as well as biostratigraphic methods, which put the evolutionary development of flora and fauna into the geological context. As a rule, a sensible combination of different methods will lead to the best results in geology, especially when it comes to investigating the age of rocks with a complex geological development history.

Chap. 2 is dedicated to the way in which rocks can store the polarity of the Earth's magnetic field. In summary, it is somewhat different with sediments than with cooling lava, but basically, for all rocks, the components of the cooling or solidifying rock that respond to the prevailing Earth's magnetic field "freeze" the orientation of the then prevailing Earth's magnetic field at a certain point in the rock's formation and store it over geological periods. For our considerations, it is important to note that equally "thick" layers often represent significantly different time durations (Fig. 4.8).

4.1 Magnetostratigraphy

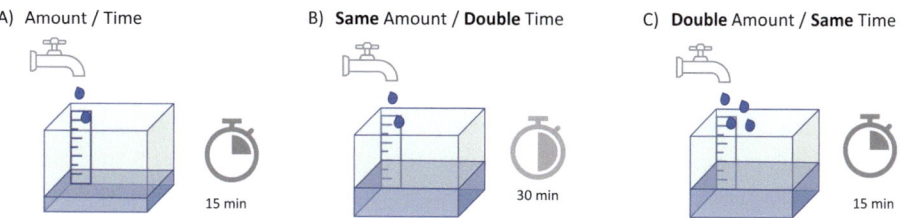

Fig. 4.9 The thickness of a rock layer, geologically speaking, corresponds to the amount of material delivered per time. The examples A to C show this in analogy to filling a water basin. A layer twice as thick results either from the same water flow in double the time (A) or double water flow in the same time (C)

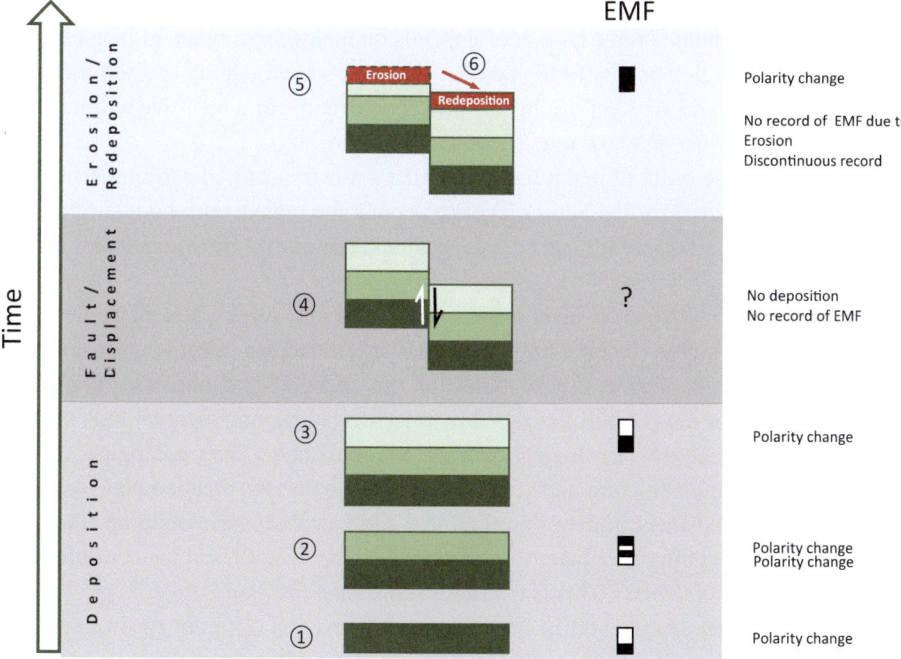

Fig. 4.10 The three rock layers ① to ③ are formed one after the other. During the deposition of layer ①, the polarity of the Earth's magnetic field changes from normal to reverse (black to white). During the deposition of layer ②, the Earth's magnetic field (EMF) changes two more times. During the deposition of layer ③, there is another change in polarity. At a later point in time ④, the three layers are offset against each other by a tectonic disturbance. Since no deposition occurs during the tectonic phase, no magnetic field can be recorded accordingly. Since the tectonics have created a relief in the Earth's surface, the higher areas are now eroded ⑤ and the debris is redeposited in the lower areas ⑥. During this deposition, the then prevailing normal EMF is recorded. The state of the EMF between the deposition of layer ③ and layer ⑥ remains unknown due to a recording gap (see also Fig. 4.4, 4.5, 4.6 and 4.7)

The polarity pattern in the figure is fictitious and serves only the purpose of symbolizing the simultaneity of the deposits at different places of the also fictitious deposition space. For simplicity's sake, we assume that the deposition at different places of the deposition space takes place at the same rate. Otherwise, for example, a white interval at one place of the deposition space would have to be represented longer/shorter than at a corresponding other place, since more/less sediment would be delivered in the same time (Fig. 4.9). For the purpose of our thought experiment (or in the figurative sense of a similar geological question), the actual, absolute duration of the (fictitious) pattern is irrelevant.

Now let's turn to Fig. 4.4. The sequence begins with the deposition of the dark-colored layer. This deposition is marked by a senior symbol, as it currently represents the oldest deposition. Let's assume a lake deposition with rivers or streams, whose introduced sediment load accumulated on the lake floor over millennia. The rate at which such suspended particles in the water, i.e., finest rock particles and small sediment grains, accumulate on the lake floor depends on many environmental factors. These include the relief in the vicinity of the lake[12], climate[13], water quantity and sediment load (also for example arid or humid climate) and much more. In short, the sedimentation speed depends on what and how much is delivered when.

For an idea of the order of magnitude, let's take a rate of 1 mm of additional sediment per year (per unit area) for the Senior layer to deposit the typical dark humus-rich clays, in our example of a lake. Building up a layer thickness of 100 m would therefore take 100,000 years.

The overlying, next younger layer Person in Fig. 4.4 may represent a different type of deposition: Let's assume the lake has silted up, the climate has become drier and more desert-like. There are several explanations for the changed environmental conditions. For example, one of the possibilities could be the changed geographical latitude and thus changed climate zone over geological periods. We remember: Over geological periods, the earth is a dynamic structure and constantly changing due to so-called plate tectonics. The typical dune deposits (sands) of our fictional geological scenario build up at a rate of 10 mm per year. Building up a layer thickness of 100 m of the Person layer would therefore take (only) 10,000 years. This is ten times faster than the Senior layer.

The next younger layer Child in Fig. 4.4 may represent a different type of sediment again. Let's assume a lagoon cut off from inflows of fresh water, slowly evaporating under a still arid climate. Let's further assume that sea water ingresses again and again into our deposition area due to changing sea levels. If the sea level drops slightly, the water is cut off and slowly evaporates (salt lake). The magnetic recording follows the principle of a chemoremanence (Sect. 3.2). The layer buildup now proceeds very slowly. It takes 1 million years to build up the next 100 m of sediment of the Child layer.

---

[12] High mountains in the surrounding area are an excellent source area (Fig. 3.16).

[13] Rock weathers differently strong/fast in warm/humid climates compared to cold/dry climates.

4.1 Magnetostratigraphy

After the deposition of the three rock layers, the earth's crust begins to stretch in a next step in the area of our former lake (Fig. 4.5). The crustal extension is due to large-scale plate tectonic movements. To compensate for this extension, the earth's crust in our example will be offset against each other along a deep-seated fault.[14]

The right slab slides relative to the left slab, so that the younger layer Child comes to lie at the same level as the older layers Person and Senior. The layer Child on the left slab now lies topographically higher than the same age layer Child on the right slab. Due to the relative elevation compared to the surroundings, no new sediment is deposited on the left slab—on the contrary, part of the upper layer Child is eroded (Fig. 4.6) and sedimented onto the relatively lower layer Child of the right slab as a new layer Baby.

Even though the layer Baby in Fig. 4.6 essentially consists of the same material as the light green layer, it has weathered, broken down into its components, namely mineral grains, and reassembled, deposited and solidified at a different time and under different environmental conditions. Baby is therefore considered a separate geological layer.

There may be a temporal gap between the youngest layer Baby and the layer Child (for the layers discussed previously, we assumed for simplicity that the deposits succeed each other without temporal delay). We do not know how long this time gap is without further investigation. One could construct a multitude of geological scenarios in which the duration of the gap could range from zero to several million years. The question would simply be when and how quickly the material of the layer Child could be eroded and relocated. This is, among other things, a function of the hardness of the rock to be eroded, the speed of movements along the fault, and thus also the regional geological development of the deposition space.

In our constructed example, we assume several polarity reversals of the Earth's magnetic field during the formation of the layers. This is illustrated by the *barcode* in Figs. 4.4 to 4.6. Note: When the originally reversely polarized layer Child of the left slab is eroded, the information about the Earth's magnetic field (Fig. 4.4, reversely polarized) is also erased. The assumption here is that the layer is eroded and transported in its individual components, rock grains and particles, to the lower right slab and deposited and solidified there again. In the course of this renewed solidification, this new layer Baby will store the polarity of the then prevailing Earth's magnetic field according to the principle of a detrital remanent magnetization (Sect. 3.2). We assume here that it is normal polarity.

To illustrate the impact of tectonic and erosive processes on the age profile of a rock sequence, let's take a look at the locations marked A to D in Figs. 4.4, 4.5 and 4.6. The respective locations are fictitious vertical drillings into the oldest to youngest rock sequence.

---

[14] Often, deformation of the earth's crust is realized by a combination of breaking rigid material and thinning similar to a long-drawn dough.

Only drilling D will core a complete profile of the rock sequence Senior to Child (Fig. 4.7). The area of drill site D is located on the right slab, whose relative subsidence along the fault to the left slab, however, has no impact in itself on the rock sequence.

The same—that is, no change in the original sequence since deposition—applies to the older sequence in the area of drilling C. However, this profile is within the catchment area of the deposition of the sequence Baby, and a drilling will bring this additional layer to light. From the description of the fictitious geological development, we know that a temporal gap between the youngest layer Baby and the older layers Senior to Child is likely.

For the further course of thought, we must note at this point that indications of a temporal gap are only sometimes recognizable to a geologist, for example due to typical sediment structures. In many cases, a rock with a temporal gap looks just like a sequence without a temporal gap. In the later dating work, especially for the paleomagnetist, caution must be exercised. If we were to date absolutely, i.e., be able to write an exact time specification to each successive centimeter, a sudden jump in the successive ages would be noticeable. In the magnetic dating considered here, only another black or white field is initially added to the results.

In the area of drilling B, it gets a bit more complicated (Fig. 4.6, 4.7). A vertical drilling will, after penetrating the layer Baby on the right slab, immediately hit the fault plane between the left and right slab, to then drill the incomplete layer Person and then the sequence Senior of the left slab. In this profile, therefore, the youngest layer Baby of the right slab will lie directly on the lower sequence of the layer Person, but with a temporal gap. In particular, the layer Child was not captured at all, and the layer Person of the left slab was not completely cored. Looking at Fig. 4.6, it becomes clear: If drilling B had been set a little further to the right, a part of the layer Child of the right slab would have been captured before the fault and then the layer Person of the left slab is captured.[15]

---

[15] We do not imagine the fault here as a layer, but as a simple thin dividing line between the sequences. In a geological description of rock sequences, a fault can also be recognized, for example due to crushed rock material, where the two massive rock packages have moved against each other and thereby crushed. How "wide" such a fault zone is or, in other words, how much material above and below the disturbance was crushed, essentially depends on the magnitude of the moved blocks and the original hardness of the rock (e.g., hard granite or soft limestone). In large fault zones, the extent can be in the tens of meters range. When dating any methodology, but here especially in magnetic dating, it is then necessary to pay special attention to the possible destruction of the age information in the rocks of the disturbance zone. Let's imagine the difficulties when typical life forms in the sequences are to be used for age dating, let's say, of the Cretaceous period—a specialized, but long-established methodology in geology, also called biostratigraphy. The Cretaceous fauna and flora preserved as fossils were mechanically crushed in the area of the fault zone and changed beyond recognition due to the high temperatures and pressures that arise in the contact zone. It is easy to see that biostratigraphy is faced with problems here. The same applies to the magnetic information stored in these sequences. Both the mechanical processing of the rock and high temperatures and pressures can affect the stored magnetization of a rock and irretrievably

In the area of borehole A, all the age information about the layers Senior to Child is no longer preserved. The upper half of the Child layer was eroded, with the information about the Earth's magnetic field being lost accordingly. The eroded material was relocated and redeposited in the area of the right fault block. In discussing borehole B, we have already addressed the re-magnetization and the temporal gap of the sediments now deposited as the Baby layer. Figure 4.10 summarizes the interplay of deposition, tectonics, and changing states of the Earth's magnetic field once again schematically, before we venture into a fictional sampling of the layers in the next section.

Sampling and dating of the fictional rock layers
The geological development of our fictional deposition space (admittedly a, albeit intuitively understandable, geological term, which may sound at least unusual in everyday, non-geoscientific language use) has led to a sequence of four rock layers. In this scenario, all layers have recorded the state of the Earth's magnetic field prevailing at the time of their formation. Since we have only assumed the deposition of sediments for our imagined scenario, the recording of the magnetic signal will have been according to the principle of depositional remanent magnetization (Chap. 3). We have not assumed any geological process that would have led to a subsequent change in the stored magnetic signal. How old are the sequences?

To perform a magnetostratigraphic dating, we essentially measure the magnetic barcode of the rock layers and compare it with the known pattern of polarity changes captured in the GPTS. The results of such paleomagnetic studies can provide surprisingly precise age information for a wide variety of rocks in Earth's history, but only if two major puzzles are solved. First, the polarity of the Earth's magnetic field recorded in the rocks at the time of the rock's formation is usually quite well hidden under all possible other magnetic signals and therefore needs to be uncovered and analyzed in a multi-stage laboratory process. The second part of this detective work lies in the fact that the geological environment during and after the storage of the magnetic signal in the examined rock can play an often complex role in the interpretation of the magnetic signal and thus ultimately also in the dating of the rock. We will look at these two "forensic" aspects (if there were a court) of our investigation in more detail below.

For our fictional magnetostratigraphic investigation, we need rock samples. As we proceed, we will better understand why, especially for direction-related magnetic investigations, the rock to be examined must actually be brought to a paleomagnetic laboratory.

---

erase it. In both cases, biostratigraphy and magnetostratigraphy, however, there is always the possibility of finding a snippet of the required age information that was not destroyed, if only one looks closely enough. Often it is advisable to combine the partial information from different dating methods and thus arrive at a result (more on this in Sect. 4.1). In our constructed example, we did not assume any extension of the disturbance into depth and accordingly represented the disturbance as a simple line.

Depending on the circumstances, i.e., the possibilities of sampling, there are several approaches to getting our rock into our laboratory in sufficient quantity. Typical paleomagnetic rock samples are small, cylindrical mini drill cores with a diameter of about 2 cm and the same height. In the most physically demanding case, our fictional terrain is only accessible on foot, and accordingly, a hand drill and other accessories must be brought on site to drill such mini cores directly from the rock.

As part of larger geological investigations, whether for research purposes such as in the IODP or for industrial purposes, for example in the context of resource exploration, larger drill cores are initially extracted with heavier equipment[16], which can then later be sampled with paleomagnetic samples. This usually happens again in at least visually less attractive core stores, where the larger cores[17] are laid out, examined, described and can be sampled for special investigations. Where the larger drill core is sampled is extremely important. How this decision is made is subject to the same criteria that must also be applied for sampling directly in the field with the hand drill. In this respect, we return at this point to the hopefully beautiful landscape of our fictional geological scenario. We undertake our virtual sampling less to demonstrate our craftsmanship, but rather to discuss another essential element in the interplay of geology and paleomagnetic sampling.

First of all, the question would be where we can generally access the buried layers. Looking at Fig. 4.4, 4.5 and 4.6 and the marked locations A to D, a common problem in geological fieldwork becomes clear: You are standing on a meadow (as an example of a plain in nature), and the interesting geological layers are directly beneath you. In the mentioned figures, a small trick is used by suggesting the possibility of larger drillings at the sample locations. The other option would be to search until you find an outcrop. Fig. 5.8 explains the formation of a geological cliff through erosion—for example when a river cuts into the ground over many thousands of years. There are numerous other ways in which rocks can be exposed in an outcrop. Every rock wall in the mountains is also a geological outcrop. Whatever the scenery, in the absence of tectonic displacement (see mountains), rock layers lie horizontally on top of each other, and for sampling we need the vertical cut. From the mentioned possibilities, we choose at will and now stand in front of our outcrop wall (Fig. 5.6 of the Bamiyan cliff may be an illustration).

In principle, we want to take our samples along a profile perpendicular to the position of the layers, i.e., along the time axis. After all, the layers have "formed" in the vertical direction. But how many such samples do we need?

When examining the ocean floor to determine the magnetic stripe pattern, we already addressed a similar problem. The samples must be taken in such a way that we do not miss any of the polarity changes. The idea of simply taking as many samples as possible, i.e., in our case, taking a sample every 2 cm according to the sample diameter of 2 cm, is

---

[16] In a suitable case, a drilling ship like the *JOIDES Resolution*, for example.

[17] Usually cut to 1 m length, measuring about 15 cm in diameter.

4.1 Magnetostratigraphy

not wrong. Unfortunately, this approach is usually not feasible, as a large number of individual samples would be needed for longer profiles—a 10 m long profile would already yield 500[18] magnetic samples. 10 m of rock wall is not a particularly long profile and may have limited geological significance. So, for magnetostratigraphic investigations, we would usually expect a much longer profile and thus many more samples.

We will see that each of these samples must undergo a complex laboratory procedure to analyze the magnetic directions. Time and money also play a role in science.

But it starts with drilling the samples in the field—500 sample points in the vertical must be determined and measured (what this means: later) and just cored with a rock drill. Of course, you can work in a team, but anyone who wants to accuse lack of effort should be reminded of the alternative sampling when sampling a larger borehole. If continuous sampling were done for paleomagnetism, there would be no material left for any other investigation (geochemistry, paleontology, sedimentology, etc.). The main drill core, brought to the surface with great effort, only provides a limited amount of material. This would not be helpful in the sense of the interdisciplinary teamwork of various geoscientific disciplines to solve geological questions, as postulated above. The actual point here is that the cost-benefit principle is also relevant in paleomagnetic investigations. Costs are not purely monetary in this context.

The solution is to take as many samples as necessary and as few samples as possible. Such a consideration involves a geological approach[19] to the layers to determine whether a rock is present that has formed rather slowly or rather quickly in geological terms. In our fictional geological excursion to the formation of layers A to D, we have encountered several examples of sediments of various origins and their different accumulation rates. Accordingly, a meter-long vertical profile contains varying amounts of time. Sometimes 1000 years, sometimes 1 million years. The geological rock approach does not give us an exact value. We only get this value with the successful dating of the rock. But ideally, we can assume an order of magnitude for the time it took to build up this "meter" of rock.

In a second step, we now have to make an assumption about how many polarity reversals are to be expected in the rock profile to be sampled. Since we want to examine the magnetic signal of the rock layers, this question may seem as nonsensical as the question about the time contained in the layers. However, in this case too, there is a possibility of approaching the answer so that we can better plan the actual investigation.

In our considerations for the creation of the GPTS, we have already discussed the average duration of the listed chrons and sub-chrons (Chap. 2). When looking at the GPTS, one also notices that some periods of Earth's history are characterized by more frequent polarity reversals than others (Fig. 4.11). For example, there is a period of about

---

[18] As always in this book, the quantity of 500 is completely arbitrary. The reader is simply meant to get the idea of "a large amount per examined section" for the discussed investigation. In principle, 500 samples can easily be analyzed in a well-equipped paleomagnetic laboratory.

[19] As mentioned, some geological expressions may sound at least unusual in everyday language.

Fig. 4.11 Schematic representation of the GPTS. The complex generation of the Earth's magnetic field in the Earth's interior is also the reason for the irregular polarity reversals in the course of Earth's history. There is no identical pattern of several polarity reversals repeated several times. In the same sense, there were always phases of millions of years of very frequent polarity reversals with durations of a few tens of thousands of years to phases in which the Earth's magnetic field remained in the same configuration of north and south poles for many millions of years. The order "frequent—occasional—infrequent" chosen in the figure is purely random and does not imply any order of such clusters in Earth's history

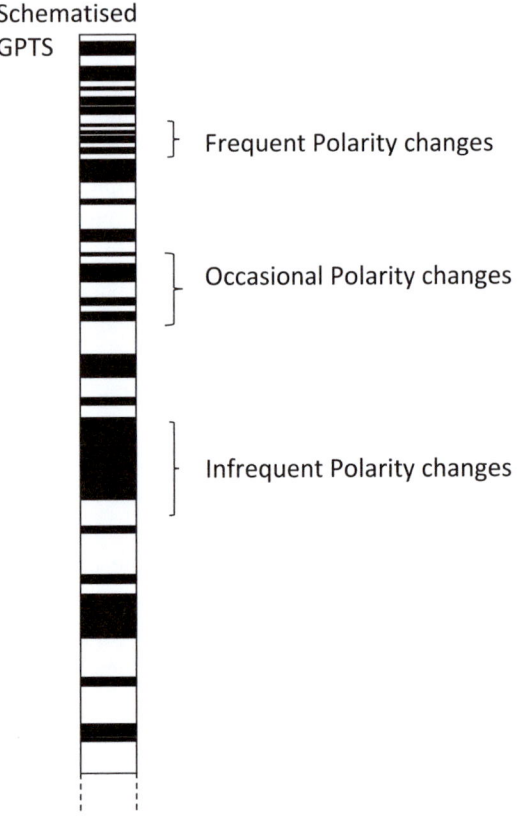

40 million years in the Cretaceous when almost no reversals from the prevailing normal polarity can be detected. A long black bar marks our magnetic barcode at this point. In paleomagnetism, this period is also referred to as the long Cretaceous quiet zone.

In contrast, several hundred polarity reversals are documented in the GPTS from the immediately preceding approximately 40 million years, especially from the younger half of the Jurassic period. The Earth's magnetic field has repeatedly flipped on the order of tens of thousands of years—i.e., in the rank of shorter sub- or even shorter crypto-chrons. The barcode of the GPTS rather resembles a sequence of thin black and white lines at this point in Earth's history.

If we recall that the Earth's magnetic field is characterized by a complex dynamo process and thus a rather unpredictable dynamics in its field generation (Chap. 2), these striking differences in the GPTS should not surprise us. We have noted that the barcode of the GPTS, precisely because of its generation in a complicated dynamo process, does not show the same pattern at any point in Earth's history (Chap. 2), since, figuratively speaking, only one "drop" is always missing to make the barrel (the Earth's magnetic field) overflow (flip).

4.1 Magnetostratigraphy

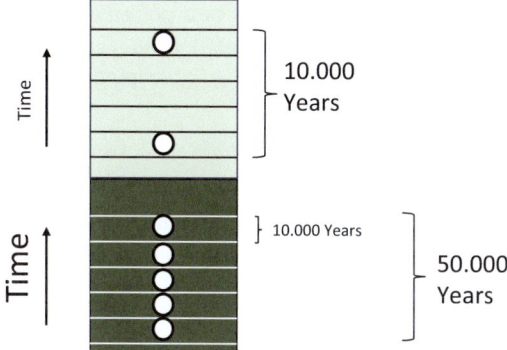

Fig. 4.12 From two rock layers of the same thickness, different numbers of magnetic samples must be taken if the two layers were formed at different rates, but the same temporal resolution is to be achieved between the samples (Fig. 4.3)

The mentioned extreme examples of the magnetic quiet zone and a contrasting Jurassic, high-frequency magnetic "techno rave" are likely manifestations of this uncertainty in the generation of the Earth's magnetic field. It seems just as likely that the Earth's magnetic field constantly flips as not at all or just now and then. For magnetostratigraphy, it is always advantageous, as clear patterns in the sequence of polarity reversals are the basis of this dating method.[20]

In summary of this somewhat longer digression to answer the initial question about the optimal distances for our samples along profiles A to D, we therefore conclude:

- In a rock deposited more quickly, there is less "time" in the direction of the emerging/existing rock than in a rock that formed more slowly. In other words, it took longer to build up the same thickness of a comparable rock layer during slow deposition (Fig. 4.9 and 4.12). A geological assessment to determine the type of rock present can provide information here.
- The average duration of polarity changes in Earth's history is about 250,000 years, but varies significantly from several million to about 30,000 years.
- The individual lengths of polarity changes in Earth's history do not follow a specific order; however, individual periods and stages of Earth's history are characterized by

---

[20] What has been said should not be interpreted as a danger that our current magnetic field could flip—so to speak, overnight. That the Earth's magnetic field will flip again in the more or less distant future is likely. If a process that ultimately results in a flip were to begin (measurable, among other things, as changes in the intensity of the Earth's magnetic field), it could take hundreds to thousands of years for this polarity reversal to be fully completed. This is obviously a short period of time only in geological terms, especially since it could still take several million years for a complete process of flipping to begin. Whether the shielding effect of the Earth's magnetic field ever completely collapses during such a process or the predominant dipole character of the field we discussed breaks up into several different—multipole—during the flip is certainly a subject of scientific discussion. The question of whether the shielding effect of the Earth's magnetic field would ever completely disappear is also discussed with this.

frequent, occasional, or rare polarity changes (Fig. 4.11). If the approximate age of the rocks can be estimated from general geological considerations or already conducted studies, a look at the GPTS reveals the frequency of polarity changes we should expect.
- If frequent polarity changes of correspondingly short duration are expected, samples must be taken at closer intervals than with few changes of long duration.
- To achieve the same temporal resolution with the same number of samples, more samples must be taken from a rock profile that was deposited over a longer period of time than from an equally thick rock package deposited in a shorter time (Fig. 4.12).
- A good strategy for sample frequency is based on a combination of both considerations.

Once we have developed a suitable sampling strategy in advance of the sampling, the samples can be taken from the rock wall or, if it is a core storage, the drill core. We can safely skip tips and tricks for the practical implementation of sampling in the context of this book, because our concern is understanding the principles. However, we still need to address a few points in this context:

- Before extraction, each sample must be measured in its spatial orientation. Usually, a compass is used for this purpose, with which at least one surface of the later sample, as it lies in the rock association, is measured. In principle, you draw an arrow on this surface and determine in which geographical direction and at what inclination to the earth's surface this arrow (and thus the surface) lies. Back in the lab, it can be traced in which spatial orientation the sample was once stored in the field. This information is essential when you consider that we intend to determine the magnetic vector of the Earth's magnetic field in this sample. And what sense would this make without knowledge of a reference system?
- For paleomagnetic studies, several samples are usually taken from the exposed rock, namely parallel samples from the same layer. Samples, therefore, which potentially, ideally, should deliver exactly the same result. This is justified in the intention to minimize inaccuracies of the determined magnetization direction.[21]

---

[21] What could be interpreted as playing with statistics or generally lacking trust in the relevance of a single magnetic measurement, in reality, has several plausible reasons. It should be noted here that our magnetic investigations aim at the most accurate determination of a direction (of the Earth's magnetic field) in a reference system (declination and inclination in relation to the geographical system). There are many possible individual inaccuracies, which can add up and directly translate as errors into the later result (of the magnetization direction). Here 1° inaccuracy when reading the compass, here 1° inaccuracy when marking the samples, here 1° when drilling the sample etc. All avoidable? Even when drilling in difficult terrain, possibly with poor accessibility of the naturally uneven rock wall and a not fundamentally handy drill, rocks are not laterally homogeneous and fatigue sets in after the 500th sample? Even if you work as accurately as possible in every respect, taking several supposedly identical samples offers the opportunity to avoid

4.1 Magnetostratigraphy

When we talk about sampling rocks, most of us will literally think of "rock hard" material. Of course we do, because they are stones, at least if we want to call the geological layers of the earth's crust so—somewhat casually. In doing so, we overlook the fact that many "stones" of the earth's crust form a stable foundation under the load of the layers above, but have not yet been fully lithified. For example, some sediments of the ocean floor (mostly clays), which are drilled from depths of several hundred meters and are accordingly already hundreds of thousands of years old, can still be pressed in with the finger—like unhardened clay when modeling a vase. The complete "fossilization" only occurs at even greater depths, higher load, higher temperatures, cementation or simply after even longer time, when water has been displaced from the finest pores between the components.

In another example (see Chap. 5), there are sediments like in the Bamiyan Valley in Afghanistan, whose grains only form a solid framework and thus a rock because they are wedged together by load/pressure and thus do not fall apart. The removal of smaller samples from the rock association (i.e., removal of the load) and/or contact with water during drilling (water penetrates into the pore spaces of the sediment and washes it up) cause this sample to disintegrate into its granular components.

And here comes the important reference to our magnetic analysis: Deforming the sample, certainly disintegrating the sample, will also affect or even completely destroy the orientation of the magnetic vector.

In our minds, we stick a pencil into a block of foam, compress this block and observe that the pencil (the magnetic vector) rotates into a flatter angle as a result (Fig. 3.4). We imagine the stored direction of the Earth's magnetic field in analogy to this pencil (or the pencil as an arrow). If the rock is deformed or even just rotated out of its original, horizontal position, the orientation of our magnetic vector, i.e., the direction of the recorded Earth's magnetic field, also changes accordingly. How is this to be understood?

Just as there is no bar magnet in the earth's interior, magnetic records in rocks do not exist as elongated objects, certainly not as pencils. Rather, each individual remanent magnetic mineral (grain), which is involved in a record of the earth's magnetic field in a

---

systematic errors in sampling (sometimes you hold the drill 1° in one direction, the next time 1° in the other direction wrong). You understand the picture and the various consequences. In the case of paleomagnetic sub-sampling of drill cores, due to the small diameter of such drill cores, often less than 15 cm, there is no other option than to have to work with only one magnetic sample per stratigraphic level. If there is the possibility of optimizing the accuracy of the result with several samples when sampling an outcrop wall, this opportunity should be used. As always, however, it is also a cost-benefit consideration here. To establish a polarity change as a new element of the GPTS, the highest precision is required and sampling with several parallel samples is certainly indispensable. If the investigation is purely aimed at magnetostratigraphic dating, it may only be important to note that negative or positive inclinations predominate in one level (normal or reverse polarity). An inaccuracy of a few (!) degrees may then be acceptable in a rather pragmatic view.

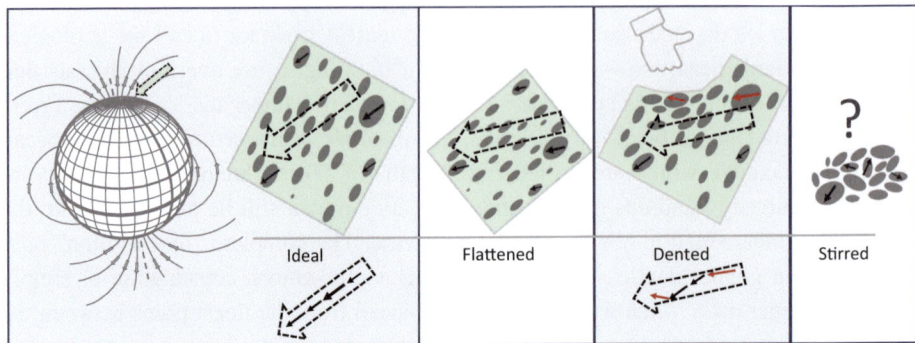

Fig. 4.13 What happens when a sample is deformed during extraction? The illustration shows, highly schematically, the alignment of individual remanent magnetic minerals (grains with arrows) along the field lines at the place of origin (globe far left) with the resulting "overall direction" as a larger dashed arrow. From left to right, four examples are shown: **a** undeformed sample (ideal), **b** flattened sample, in which all minerals are rotated in the same way, **c** indented sample, in which the original alignment of individual grains is individually rotated, **d** destroyed sample, with loss of the original grain composite. Below **a** and **c**, the overall direction as a result of the individual directions of the grains is exemplified (vector addition; Fig. 4.14 ff.). Even if the storage of the magnetization direction does not necessarily rotate the magnetic grain in the direction of the earth's magnetic field lines (Chap. 3), but this magnetization occurs internally, from the moment of storage any movement of such a grain will also result in the movement of the magnetization arrow

sample, contributes "its stored direction" to the overall result (Fig. 4.13). To the result, which is measured with a magnetometer. Just like individual small magnets, whose magnetization direction we think of as an arrow (or somewhat nonchalantly as a pencil).

In an ideal record of the earth's magnetic field, each of the involved grains (each of the arrows) points in the direction of the recorded earth's magnetic field. We can—in this sense—imagine each of the involved remanent magnetic minerals as a small magnet. When the sample is deformed, the grains change their position relative to each other.

In the simplest case, the grains move closer together due to increased overburden, by reducing existing pore space between the grains. If the sample were to be pressed in with the thumb during extraction, for example, the individual grains would change their relative position to each other within the rock. In this case, the measured "overall direction" would also change, as the involved grains no longer point in the same direction due to individual rotation. In Fig. 4.13, various scenarios of deformation and the corresponding impact on the resulting magnetization vector are discussed. Even in this illustration, the sizes and quantities of the remanent magnetic minerals in a rock sample are dramatically simplified and exaggerated. If all components of rock deformation were known, such a direction distortion could theoretically/mathematically be corrected.

If a rock sample is indented or flattened during sampling, it would be highly unrealistic to attempt such a correction, as it would be unrealistic to reconstruct the relative displacement of each individual grain. As noted earlier, we should not be deceived by the

4.1 Magnetostratigraphy

schematic representation in Fig. 4.13. In reality, both relative grain size and number of grains vary.

If the rock completely disintegrates into its components due to erosion, the individual grains, outside their original composite, point in random directions in the worst case. The original direction of the stored earth's magnetic field cannot be reconstructed in either case of deformation and the corresponding sample is useless for magnetostratigraphic purposes.

Once we have collected the desired number of samples in the field based on our preliminary plans, these rock samples are taken to a suitable paleomagnetic laboratory for further investigations. There are numerous such well-equipped laboratories worldwide. All reputable laboratories are usually located at universities and affiliated research institutions. At this point in our virtual investigation, we simply note that, unlike rock magnetic and earth's magnetic field-independent studies (Chap. 5), there is hardly any possibility to perform the measurement of magnetic directions directly in the field. Sensitive magnetometers are necessary, and in particular the usual challenge of revealing the "true" direction of the past earth's magnetic field from a multitude of other states of the earth's magnetic field stored in the sample. Even though one might first think of using a hammer when "exposing" in a geological context, various and quite sensitive so-called thermal and electrical demagnetization methods are actually available for paleomagnetic analyses. The expected final result in a magnetostratigraphic investigation will be a direction indication for each sample, i.e., declination and inclination of a magnetization direction.

We do not want to illuminate the laboratory experiments necessary for each magnetostratigraphic investigation in detail here. At this point, it should only be emphasized that the so-called sample preparation and the multiple measurement of the individual samples with a suitable—i.e., shielded against external magnetic interference fields—magnetometer can take several days depending on the laboratory equipment and of course the quantity of samples.

When individual samples are measured, they must be brought to an exact size and shape in order to be measured in the sometimes space-consuming magnetometers. Many laboratories now ease the efforts of the investigators through automated measurement procedures, in which, for example, a robotic arm places the samples one after the other into the magnetometer and starts the repetitive measurements. This step—the so-called demagnetization[22]—must be carried out step by step for each sample, and

---

[22]The idea is that different remanent magnetic minerals in a sample have recorded further states of the EMF even after the time of rock formation. Not all sizes and types of remanent magnetic minerals in a sample are ideal recorders and sufficiently stable. The initially measured magnetic direction is therefore often an "apparent" direction because it results from the superposition of many individual directions. The successive reduction of the individual directions in the laboratory is referred to as demagnetization and is translated into a (limping) analogy in Fig. 4.14.

the measurement can, at least if no robotic arm is available, sometimes last "all night". However, disturbing magnetic fields are often restricted at night (traffic, power grid, etc.), which in turn can be good for the results of the magnetic measurements.

The result of the measurements is always a data sheet full of numbers. One could argue that at this point in the investigations, the actual, substantial part of the work is only just beginning.

From the data sheet, the "numbers" must now be determined for each individual sample, from which the direction of the Earth's magnetic field at the time of the rock's deposition can be calculated. Special graphics (usually so-called orthogonal vector diagrams) are often used for this purpose to select the "correct" direction from the existing directions in a sample. As a non-magnetic analogy, let's consider the following scenario.

> Figs. 4.14, 4.15 and 4.16 illustrate this part of the magnetic analysis using a journey from Cologne to New York. As always in this book, we construct an analogy, and as usual, this thought experiment could have many variations. The only goal is to better understand magnetism through everyday examples.
>
> In magnetic investigations, the original magnetization direction of the EMF is usually sought—the magnetization that was acquired at the time of the rock's formation. In our travel analogy, this would correspond to the search for a possible first

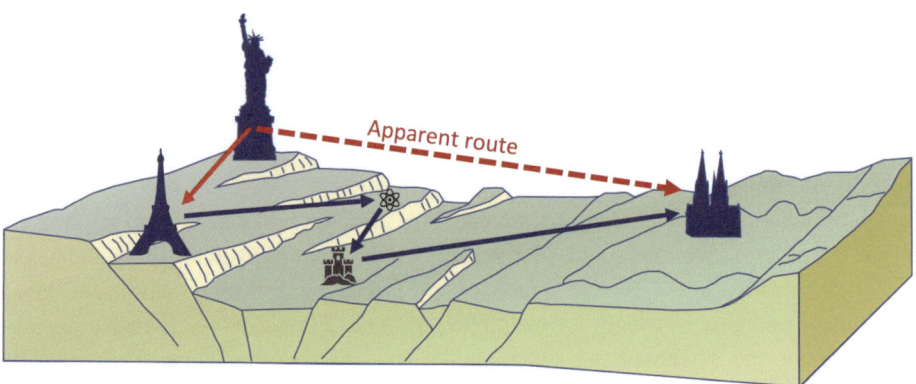

Fig. 4.14 In analogy to demagnetization attempts and vector addition in paleomagnetic studies, the reconstruction of a travel route from New York to Cologne is considered here. In both cases, it is to be determined in this sense whether a journey/magnetization took place directly or whether detours and stopovers were made. Mathematically, this is a so-called vector addition, where an arrow simply points directly from a starting point to an end point and thus represents the sum of possible detours, but does not make clear which detours were made. Detours and stopovers of our journey thus symbolize the recording of later polarities of the EMF since the formation of the rock. The first magnetic recording from this formation time is overlaid by the directions acquired later. The sum of the magnetization directions stored in the rock is always measured in a magnetometer. The individual directions must be reconstructed step by step

4.1 Magnetostratigraphy

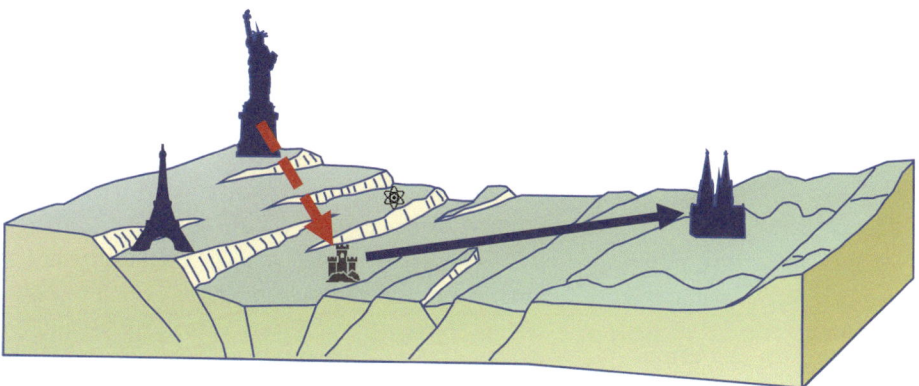

Fig. 4.15 Illustration of the second stage in the reconstruction of a journey/the overlay of several magnetizations (see also Fig. 4.14 and 4.16)

Fig. 4.16 Illustration of the third stage in the reconstruction of a journey/the overlay of several magnetizations (see also Fig. 4.14 and 4.15). Strictly speaking from a natural science perspective, the described routes are the analysis of a vector addition. In Chap. 3 different reasons are given why magnetic minerals in **one** rock sample may record different states of the Earth's magnetic field at different times. Different remanent magnetic minerals (or simply different sized remanent magnetic minerals) are predictably recognizable by different methods—just like in the described journey

stopover on the journey. However, we do not know if such a stop was made. We only know that the journey started in New York and ended in Cologne.

So let's assume a person, object XY, is currently in Cologne and was previously in New York. In Fig. 4.14, this is represented by the dashed line labeled "apparent path". In our analogy, this apparent path from New York to Cologne represents the measured direction of the so-called natural remanent magnetization (NRM). This is, so to speak, the magnetization arrow, which is determined with a magnetometer when a rock sample is measured for the first time after being taken in the field.

In a magnetic demagnetization in the laboratory, we now apply the first, mild steps to see if there were deviations from this path. It is checked whether slightly increased temperatures, for example, lead to a change in the direction of the arrow. In our journey, these are the first simple thoughts about the journey.

The statement "originating from" is based, for example, on a detectable distance and a direction pointing to New York. In the case of a work of art, this could be the creating artist, for example, who demonstrably never left New York in her lifetime. When the departure/shipping of the work of art from New York took place is not known in detail. However, if the artist or a certain creative period can be named, a maximum age of shipping is given. Similarly, in the magnetic investigation of a rock, the formation period for the examined rock can often be narrowed down, for example to "not older than …". There are numerous reasons for such geological statements, from the supra-regional classification of a study area to the successful dating of rocks in the local context.

The direct path from the starting point to the end point of the journey may be assumed as the route as long as no other information is available.

It is important: This direct path may never have been taken, hence the designation "apparent" in Fig. 4.14.

In the example of the journey, intermediate stops along the way only become evident when we look more closely. The easiest way to determine recently passed stations is because, for example, documents and other travel expenses have not yet been recycled. A gallery expose from "Burg" dated shortly before today thus confirms a stop before reaching Cologne. There are no indications of other journeys in the period between Burg and Cologne; this route is therefore considered secure, both in terms of direction and travel distance and date. In determining the magnetic directions, we have just carried out a first demagnetization step and "destroyed" the last (youngest) of the overlapping magnetizations. The now measured resulting magnetic direction corresponds to the route from New York to Burg (Fig. 4.15), the reduced (magnetic) direction of the route from Cologne to Burg. Whether the journey ever took place from New York to Burg or whether this route must also be considered "apparent" can only be determined by the presence or absence of further evidence. In the case of magnetization directions, a higher level of demagnetization is experimentally driven at this point. If the direction of the resulting magnetization changes afterwards, this would be an indication of another stopover and in terms of magnetization a hint at the recording of another, earlier configuration of the EMF. In Fig. 4.14, 4.15 and 4.16 we have indeed assumed another stopover (atomic symbol).

4.1 Magnetostratigraphy

Since the principle is now clear, we do not want to construct any more reasons and evidence for another stopover before arriving in Burg for our art object. However, we can certainly draw the next resulting arrow, which we would mark as the apparent route from New York to "Atom".

The probability of reaching a destination only via detours may increase with the time span between departure and final arrival, especially if the duration is on the order of several years or—as in geological reality—many millions of years and correspondingly many different configurations of the EMF and possible geological events since the formation of the rock.

A look at Fig. 4.16 suggests in a next step that the original first journey across the Atlantic from NYC to Paris took place—the time and direction of the journey of the art object from the place of origin are thus determined. Speaking magnetically, no further overlapping directions of the polarity of the EMF at the time of rock formation can be detected analogous to the route from New York to Paris. In other words, stronger demagnetization steps no longer change the direction of the measured magnetization. This magnetization direction corresponds to the magnetization of the rock at the time of rock formation, in sediments the time of solidification after deposition or in magmatic rocks the time of cooling of the hot lava.

Anyone who has worked through Fig. 4.14, 4.15 and 4.16 may suspect that—once known—the "journey" after the analysis of the first sample no longer presents any particular surprise, since the true path/the sought magnetization direction has now been explored. However, this is only partially true.

If we look at samples taken close together (temporal, type of formation) from the same layer, i.e., geologically similar rock, the path is indeed possibly quite similar. For example, if we examine samples from our Senior layer in Fig. 4.4, we can expect a certain similarity in the "travel route".

In a sense, everyone departed at the same time and on the same train, and arrived at the same time. The usual geological heterogeneity—in the case of the Senior layer, slight, possibly only seasonal variations in the sediment load of our lake's tributaries—also means a change in the composition of the magnetic minerals in the respective samples.[23]

---

[23] As we have seen in Chapter 3, even a minimal change in the prevailing grain size of the magnetic minerals can change resistance of the minerals to external magnetic fields. This could mean, for example, that the percentage of magnetic minerals whose stored magnetic direction, which is overwritten by the drilling process, changes significantly. In the translation into our analogy from Fig. 4.14, this would correspond to a changed stopover and thus a changed route from Cologne to New York. To stay with the example of our train journey: Even if all passengers travel on the same train from A to B, the descriptions of the journey will differ. One had a relaxing window seat, another had to stand crowded in the aisle, the third didn't like his seat neighbors, etc.

Even greater differences can be expected in the analysis of the demagnetization attempts of samples from different layers and different ages. The method/principle of analysis remains the same, however, in the language of our analogy, we would have to decipher a journey from Minneapolis to Grasse (or whatever places may spontaneously come to mind) if we were to examine the Child layer instead of the Senior layer. For these reasons, each individual sample will need to be given time and attention.

The next step of our virtual analysis: As soon as we are convinced of the quality and nature of our determined directions of the past Earth's magnetic field per sample (and can scientifically justify this), we create the barcode. For this, each sample with a positive direction is marked black (on paper, not the actual sample), and each sample with a negative inclination is marked white (Chapter 2).[24] The order on our paper is created, and thus the barcode, by entering all samples according to their respective position in the rock wall, i.e., according to their centimeter specification, in our graphic (Fig. 4.12).[25]

So now we have an acceptable barcode for our sampled profile, or in other words, a pattern of thinner and thicker black and white stripes. The last step of our magnetostratigraphic dating is thus simple. We look in the standard GPTS for this pattern—for one, even when a long profile in a geological study is considered, usually very short section of a few polarity changes listed in the GPTS.

Once we have found the corresponding pattern, we can read the age of our sampled rock profile from the dated GPTS (Fig. 4.16).

A look at our fictional scenario and the drillings A to D (Fig. 4.6) shows another peculiarity of our barcode. It is of crucial importance that the barcode represents a continuous profile. A comparison of the fictional magnetostratigraphy of profiles A to D in Fig. 4.6 and in Fig. 4.16 illustrates this.

Gaps in the deposition create a barcode that is only readable in the intervening areas, as if one were actually trying to cut together a barcode on the supermarket packaging from several different barcodes. This would probably not be readable. In magnetostratigraphic practice, this means that for the layers below and above the geological gap, an

---

[24] Black corresponds to an arrangement of the Earth's magnetic poles in the past, which corresponds to today's arrangement. White, on the other hand, indicates a pole reversal in Earth's history.

[25] The boundary between black and white will be drawn halfway between two samples with the corresponding measurement results. Each of the two samples is then the last secured result of a state of the Earth's magnetic field and the boundary drawn in the middle is a "fair" interpolation with regard to the state of knowledge. Problems can arise here with a very wide initial sampling of a geological profile. After all, the width of the black and white stripes defines the appearance (and thus the information) of our barcode. If the black sample actually lies on the border and all further samples from the "unsampled no man's land" between the samples were actually white, both stripes would have a wrong width. In case of doubt, the boundary must be verified by a re-sampling, i.e., refined sample distances at this point.

4.1 Magnetostratigraphy

independent pattern in the barcode must be recognized and correlated with the corresponding section of the GPTS.

We have already discussed that the duration of the gap depends on the geological conditions of the deposition space and can well be several million years long. In the search for the appropriate place in the GPTS, one will in all cases include the known geological clues to find the appropriate place.[26] An important point to emphasize here: The pattern comparison of the measured magnetic barcode with the GPTS will usually only be possible with the help of a geological examination of the sampled profile and the entire geological situation.

Another peculiarity is shown in Fig. 4.17. Dashed lines, which connect the three depicted drillings with each other and with the schematized GPTS, are timelines. Just like the continuously drawn line marked as a timeline in the upper part of the profiles. A timeline connects points of the same age in the profiles. Anyone who follows the timelines will find that differently thick white or black areas in the drillings were connected with each other as if a connection between different barcodes had been postulated.

At this point, another geological consideration comes into focus: the already mentioned sedimentation rate (Fig. 4.8). And with this consideration of the possible stretching and elongation of our barcodes, the question of the validity of a pattern comparison with the GPTS for dating a rock sequence in space immediately arises. After all, the code consists of a sequence of black and white stripes, and a pattern emerges from the relative thickness of the respective stripes to each other. If the relative thickness could be arbitrarily changed at any point due to changes in the sedimentation rate—due to the unknown size of the sedimentation rate—any pattern from these stripes can also be generated at any point of the GPTS[27].

---

[26] For example, the upper piece of the sampled profile is younger than the section below the geological gap, unless a tectonic event would have pushed the older layers over the younger succession.

[27] It should be considered that the GPTS assumes the same accumulation rate for each of the several hundred polarity reversals—the black and white stripes—in its temporal classification. In other words, a stripe representing a duration of one million years is exactly twice as wide as a stripe representing half the duration of a polarity interval of 500,000 years. This type of normalization to a common time axis is not present in our determined barcodes from the drillings A to D of the fictional scenario (see non-parallel timelines in Fig. 4.7). The samples were indeed taken at the same or at least known distance from the drill cores, but as discussed above, 1 m of rock from different deposition conditions—possibly—represents significantly different periods of time. And this results in the shifting timelines in Fig. 4.17 when correlating the drillings. Any subsequent derivation of a temporal classification within a sampled sequence (the subsequent determination of a sedimentation rate) would then certainly be a circular argument.

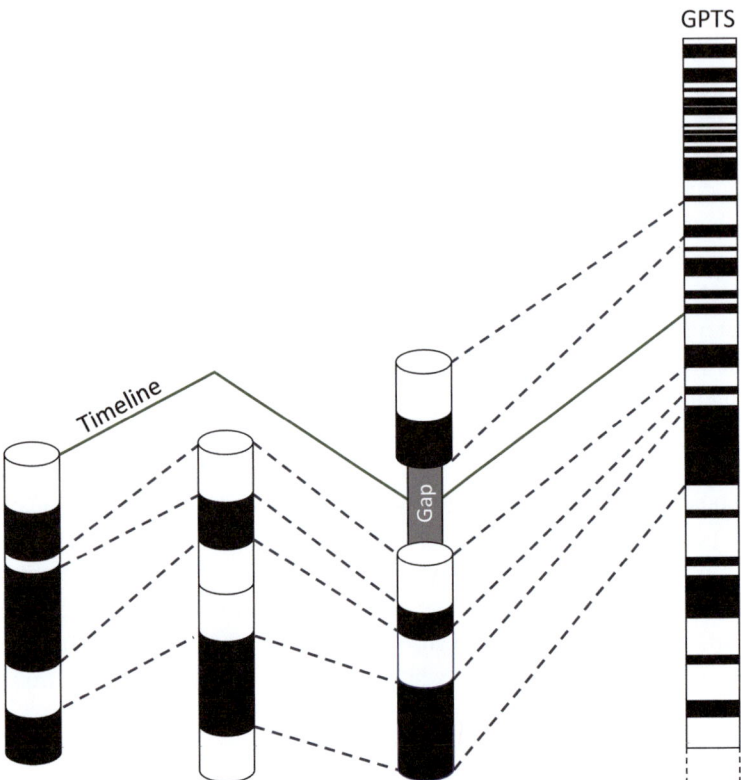

Fig. 4.17 We have created magnetic profiles at different locations of our fictional deposition space. These profiles are correlated with each other, and their pattern is matched with the pattern of the GPTS. The seemingly simple step in dating these sequences must be accompanied by a geological analysis of the rocks to recognize possible stretches and compressions of the determined barcodes. The text explains how

The solution to this dilemma lies again in geology or initially in the mental realization that the purely technical determination of magnetic directions in a rock sample will not be sufficient in many cases to make a significant geoscientific statement, i.e., to date the rock layers in the context of this chapter.

The results must be assessed and presented in the context of the respective sampled rocks and the development of the respective deposition space. Only then does the relative width of the stripes to each other and thus a pattern correlatable to the GPTS emerge.

Here are a few considerations, so to speak as cornerstones in the complex world of possible geological scenarios, to make the statement of the preceding sentence appear somewhat less nebulous: An ideal magnetic barcode consists of stripes, whose relative thickness represent equal time units/the black and white stripes do not appear relatively compressed or elongated to each other. In practice, therefore, various questions and especially the different processes in rock formation must be taken into account.

4.1 Magnetostratigraphy

Polarity reversals, for example, are not bound to layer boundaries (Fig. 4.7), that is, the relative width of the stripes in a layer, in which the geological investigation does not require a significant change in the deposition conditions, can be considered reliable. In our fictional scenario (Fig. 4.4, 4.5 and 4.6), for example, one would not have to assume large fluctuations in the accumulation rate for the deposits in a calm lake environment. It is irrelevant how high this accumulation rate (in concrete numbers) actually is. What is important is to be able to reasonably assume that the rate did not fluctuate significantly during the deposition of the rock, i.e., it remained relatively the same in successive layers.

- As in the drillings A to D, the pattern can possibly be assembled from parts of several drillings, in which the aforementioned condition is met.
- If a drilling is sunk into the sedimentary cover of the ocean floor or a lake, the uppermost layers will have been magnetized in the currently prevailing normal polarity of the Earth's magnetic field. Measurements are made until the first and the subsequent inverse samples indicate the beginning of the Matuyama Chron. If there is no tectonic disturbance of the deeper layers, the following polarity reversals can be worked through from top to bottom, so to speak.
- In the same line of thought, a stretched and compressed barcode can also be normalized using the GPTS if at least one normal or inverse polarity can be reliably—at any point—hooked into the GPTS. Provided there is no layer gap, the remaining stripes can then be worked through upwards or downwards.

A typical example of the aforementioned scenario would be knowing the age of a layer or a short section of the sampled profile from further geological investigation of the deposition space. Often it is known that a certain layer XY is sampled, which could already be dated at a possibly different location. From a magnetostratigraphic point of view, it is sometimes enough to know that part of the sampled profile belongs to the XY epoch of the geological time scale (Chap. 2). Perhaps through a specific fossil or another way to estimate the age. If there are only a few polarity changes in this XY epoch, the degrees of freedom for correlation are already significantly restricted, and the "prospect of success" is increased. The crucial point is that in this way, the hitherto undated layers below or above the known location can now also be dated and their respective durations determined through correlation with the GPTS.

In the extreme case that no knowledge of the geology of the layers, their accumulation rate, or even their approximate age is known, a purely magnetostratigraphic dating is indeed difficult. In such a case, it would only be possible to assume that there is no stretching or compression of the stripes of the determined barcode from a sampled rock profile, and accordingly to undertake a pattern comparison with the GPTS. Often this can lead in a positive sense to an iterative narrowing down of the true age of the rock layers, when other dating methods or geological considerations are also restricted in their degrees of freedom in this way. It's somewhat like having several candidates available for a missing puzzle piece and trying them out one after the other.

In the best case, to achieve an independent magnetostratigraphic correlation with the GPTS, at least three to four polarity changes without gaps per profile should be present. With a smaller number of stripes (only one black and one white stripe, for example), no clear pattern emerges even with significantly different durations of the polarity intervals.

We initially postulated that the GPTS does not display a self-similar pattern of the barcode at any point if we want to consider the sequence of differently long polarity intervals as a pattern. Such a pattern must therefore usually consist of several stripes in order for a statement to be made correctly. Against the background of a possibly compressed or stretched magnetic barcode when examining a geological profile, we now understand this statement better. But we also recognize that in the practical application of geoscientific methods, including paleomagnetism, it is important to draw conclusions only on the basis of complex geological facts.

If it has now been successful to correlate the entire barcode of a sampled geological profile with the GPTS, the actual dating of the sequence is trivial. Each of the polarity changes recorded in the GPTS has two readable "absolute" age indications: 1) how many (millions of) years ago this respective state of the Earth's magnetic field began and 2) how many years ago the polarity change then occurred. In Chap. 2 we have already discussed this property of the GPTS. Under the restriction of the above discussion to avoid circular reasoning, the actual accumulation rate for the sampled layers can then also be calculated for each of these sections (layer thickness divided by duration).

Here we come to a final step in the methodology of magnetostratigraphy: the geological sanity check—an invented term, admittedly, because of course all steps of our investigation were initiated and carried out sensibly. But—probably not only—in geology, even the best indications can ultimately lead to an unreasonable result. And we want to avoid that at all costs.

An example should illustrate this to us. In the previous discussions about the importance of the accumulation rate for assessing the stripe thickness or for determining the sampling rate along a geological profile, we cited the fundamentally "low" rates of lake sediments as examples. In contrast to other types of sediment, it takes much longer in these sediments to build up a layer of a certain thickness, as little and very fine-grained material is delivered. This slowness cannot be specified in a single absolute accumulation rate, but at most in a range of such rates.

Such a range indicates for the most diverse scenarios in the deposition of lake sediments that in the slowest cases it is so much new sediment per year, but in other cases it can also reach so much material per year. In practice, this so much can certainly be translated into a duration of 100,000 years or 1 million years[28], to accumulate a new layer of sediment at the bottom of a lake.

---

[28] As always, we only mention the concrete numbers here for illustrative purposes. What is meant is "approx. a factor of 10".

4.1 Magnetostratigraphy

This is a significant difference, but we have already seen in several examples that even small and seemingly insignificant variations can have significant effects over long periods of time. In other words, even with fundamentally "low" rates of lake sediments in our example, the range of known rates can result in a layer thickness of 10 cm or 1 m after 1 million years.[29] With the dating of the sequences, we can now calculate a specific accumulation rate for the sediments examined in our fictional deposition space.

So what is the reasonableness test? Certainly, it would not be reasonable to assume a correct dating if the now calculated accumulation rate is significantly outside the expected range of such rates for lake sediments of various provenance. If the geological description of the layers clearly implies the formation of the sediment in a calm lake, but the dating yields accumulation rates on the scale of a mountain landslide (obviously a lot of material in a very short time), the investigation must, so to speak, go back to the drawing board.

As a rule, this means in a first step the renewed search for an alternative pattern in the GPTS with then other implications for the duration of individual layers—which then hopefully also yield an acceptable accumulation rate for our fictional lake sediments.

If from the beginning a sufficient number of usable geological facts (we have given various examples: from the possibility of clearly determining the sediment type, to the classification of the deposition space in the supra-regional tectonic development) were available in the magnetostratigraphic investigation and could be used sensibly, such a re-examination will not be necessary. However, just as a detective in some particularly tricky cases can only find clues, but no definitive evidence, one will also find scenarios in geology where one reasonably follows the clues in one direction, but then ends up in the thicket (here a non-compatible accumulation rate) and has to find an alternative path (an alternative pattern match of the magnetic barcode). In some cases, the geological facts are sometimes simply too scarce due to a cost-benefit consideration.

If we are surprised that natural rocks—from the rugged cliffs of the mountains to the almost still malleable layers in a clay pit—can store the magnetic properties of the Earth's magnetic field, we will be less surprised to hear that the same rock can record the changing states of the EMF several times over its usually millions of years of existence. The few components of a rock that are responsible for recording the EMF are not all the same in their magnetic properties or, better said, in their abilities to retain magnetic records. Some of the magnetic minerals are indeed reliable recorders, and it is very difficult to erase a once stored magnetic information (Chap. 3). To do this, the rock would usually have to be melted, i.e., exposed to very high temperatures of 500–600 °C.

For example, during the collision of two tectonic plates and the associated metamorphosis of the rocks in a mountain building process, such heating or even melting of rocks can occur. More often, rocks are sunk deeper into the Earth's crust over millions

---

[29] As always, we only mention the concrete numbers here for illustrative purposes.

of years, being increasingly buried by newly deposited sediments. In this way, the rocks enter areas of the Earth's crust with temperatures higher than at the surface. The so-called geothermal gradient of the Earth indicates a warming of about 3 °C per 100 m depth away from the geologically active zones of the Earth. So we expect about 300 °C at a depth of 10 km! This is good for the potential use of thermal energy from the depths, but unfortunately disadvantageous for the permanent recording of magnetic properties, at least for some of the recording minerals.

4.2 Moving Continents with Stored Earth Magnetic Fields

In a striking, but quite instructive case of paleomagnetic consideration of rocks, one could therefore imagine that each rock of the Earth's crust is inscribed with a directional arrow at its formation. This directional arrow just needs to be made visible with magnetic methods. We have learned that this arrow not only changes its orientation according to the pole reversals in Earth's history, but also always has an angle typical for the place of rock formation, which is connected with the geographical latitude. This information can be read out, regardless of whether the North Pole and South Pole are in one or the other configuration of the Earth's magnetic field. This inscribed arrow in the rocks is usable for a number of other geoscientific applications besides magnetostratigraphic dating.

We have touched on plate tectonics[30] at several points in this book, among other things to provide the necessary basis for understanding the geological dynamics of our planet for reading this book, but as friends of the discipline of course also to highlight the lasting importance of paleomagnetism for the scientific establishment of plate tectonics.

To understand what contribution an imaginary magnetic directional arrow in rocks can make to plate tectonic work, we start with another, somewhat striking clarification: To understand the dynamics and changing development of our Earth's crust in Earth's history, it is not sufficient for modern plate tectonic models to "move" the seven large current continental plates of the Earth "against each other". For a scientifically meaningful investigation of the development of the Earth's crust, one must zoom in much more into the detail of individual crustal parts. The smallest crustal pieces in plate tectonic models are sometimes more on the order of hundreds of kilometers than in the dimensions of the continents we know. This is scientifically interesting, but also for example for raw material exploration, as the treasures of our Earth are always formed under special, local geological conditions.

One understands the necessity for more detail better with the knowledge that even the mentioned large continents like Africa or South America were also composed of smaller crustal parts (convergence) over the course of many hundreds of millions of years of

---

[30] Specifically on this topic see Wolfgang Frisch, Martin Meschede, and Ronald C. Blakey 2010, *Plate Tectonics: Continental Drift and Mountain Building*, 220 pages, Springer Verlag.

4.2 Moving Continents with Stored Earth Magnetic Fields

Earth's history, crustal parts were deformed or metamorphically changed, broke apart again at other places (divergence) and were reassembled differently, supplemented by intrusive magma, eroded and relocated etc.—until the current outlines of the continents and their relief manifested.

However, one should not fall into the idea that all continents are built from a swarm of smallest, also currently independent crustal parts; modern plate tectonic models nevertheless work with crust fragments of small extent (hundred kilometers) to reconstruct the fracture surfaces and "healed" seams of much earlier plate movements. Even if the "current" outlines of some continents have not been fundamentally changed for perhaps 100 million years, the internal structure of a continent is the key to its history. 100 million years are not a particularly long period of time in geological terms (if this has not been mentioned yet). The formation of some raw materials also often dates back further, and for this reason alone it is worth looking back so far.

Anyone looking for an illustration of plate tectonic boundaries of the Earth's crust will find a division of the globe into the Eurasian, North American, South American, Australian, African, Antarctic and Pacific Plate in an overview representation. Upon closer inspection, one will notice in addition to these seven major plates about the same number of smaller plates, for example the Indian and Caribbean Plate, as well as a certain number of so-called microplates. It is found that the major plates naturally do not correspond to the outline of, for example, Australia—as we perceive it as mainland on a world map—but, so to speak as an extended edge, also include parts of the surrounding oceanic crust. This can be understood when looking at the boundary between the South American and the African Plate, which runs through the middle of the Atlantic Ocean. This "boundary" is a mid-ocean ridge (MOR; Fig. 4.18), as we have already used it in the previous discussion in this chapter (see also Fig. 2.12, 2.13, 2.14 and 2.15). South America and Africa were once a contiguous continental mass (as part of the supercontinent Pangaea), which moved further and further apart from each other through crustal extension and intrusive magma with the opening of the Atlantic since the late Triassic (Fig. 2.16). Our already discussed oceanic stripe pattern in the Atlantic was created in this way. The one "half" of the newly formed Atlantic oceanic crust is geologically attributed to Africa, the other "half" to South America. Accordingly, the oceanic "edge" is to be understood around the outlines of the continents more familiar to us on a world map (Fig. 4.18).

Back to our actual topic: How does a stored direction in rocks help to reconstruct the position of individual crustal parts in Earth's history?

Let's stick to the overview consideration of plate tectonic movements of the Earth's crust. Let's stick to the African Plate as an example (Fig. 4.18).

We take a sample from a rock layer (we now know a number of samples) to investigate their stored magnetic directions from the time of the formation of this rock. We take these samples at a specific location in, let's say, Africa. For our thought experiment, it is irrelevant which exact location we imagine for this sampling. However, it is important that we can precisely name this place, especially on which latitude we want to take the sample. In reality, a glance at a map or the smartphone is sufficient for this.

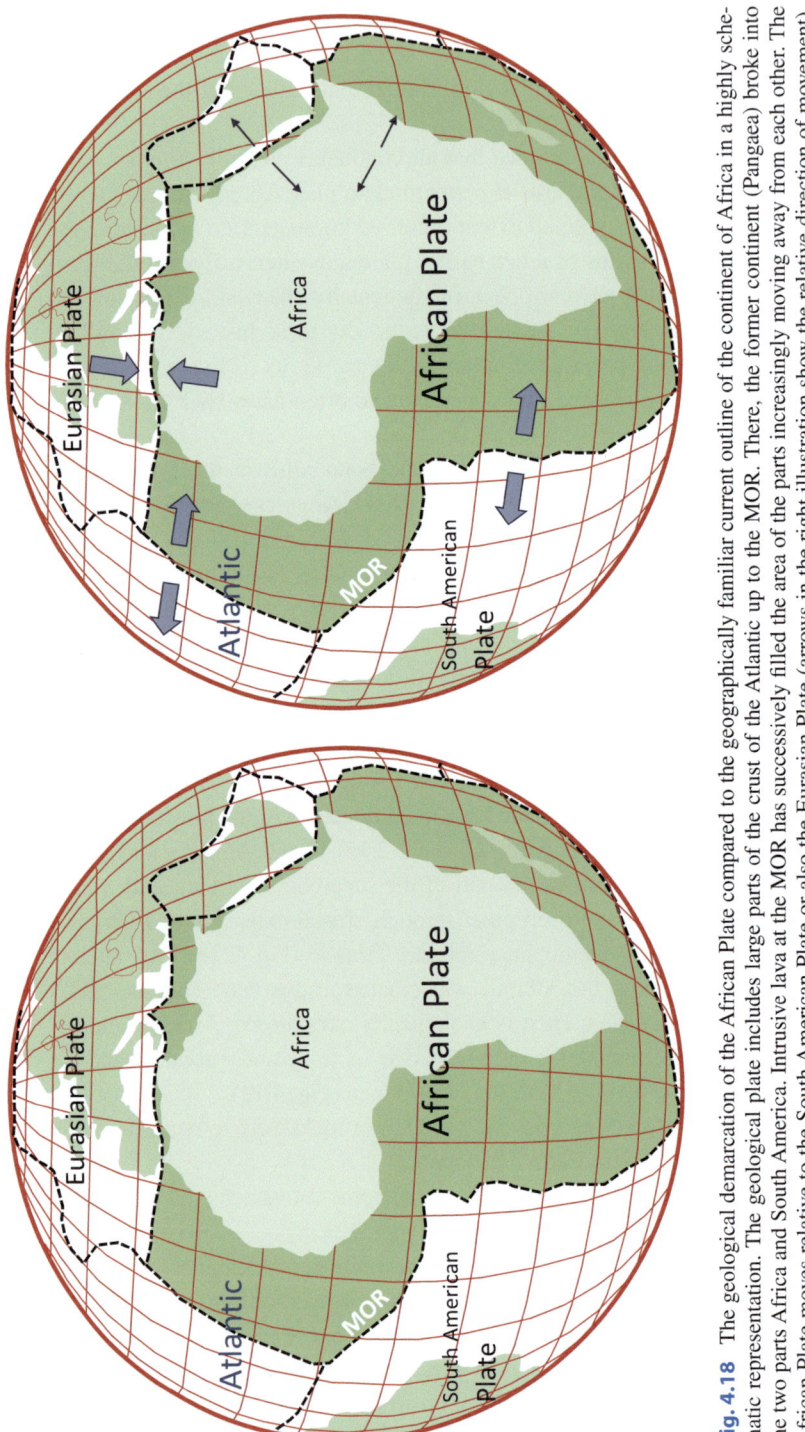

Fig. 4.18 The geological demarcation of the African Plate compared to the geographically familiar current outline of the continent of Africa in a highly schematic representation. The geological plate includes large parts of the crust of the Atlantic up to the MOR. There, the former continent (Pangaea) broke into the two parts Africa and South America. Intrusive lava at the MOR has successively filled the area of the parts increasingly moving away from each other. The African Plate moves relative to the South American Plate or also the Eurasian Plate (arrows in the right illustration show the relative direction of movement). Smaller earth plates as well as numerous microplates (with the exception of the Arabian Plate east of Africa not shown here) supplement the number of the seven main plates of the Earth's crust. This is a rough look at the "current" geological situation. Due to the ongoing geodynamic development of the Earth's crust over the entire history of the Earth, even the main plates had to be defined differently again and again, so the African Plate, for example, is to be subdivided into many further geological units. Detailed plate tectonic models, which are supposed to trace the movements of the Earth's plates since the Cambrian, therefore make it necessary to look at the Earth's crust in the smallest geological units. Paleomagnetism helps in the investigation of how such geological units (crustal pieces with an extent of a few hundred kilometers) have developed in the history of the Earth. In the text we discuss the use of the "magnetic direction arrow," "inscribed" into the rocks since the respective rock formation. The importance of paleomagnetism for the reconstruction of plate movements in the history of the Earth is also highlighted

4.2 Moving Continents with Stored Earth Magnetic Fields

Furthermore, we assume that we know the age of the rock from which we take the samples. Whether the dating of the rock was done via a magnetostratigraphic dating, another non-magnetic dating method, or a combination of several dating methods (preferred solution; Sect. 4.1), is irrelevant for our consideration. Important for our thought experiment is that we sample dated rock layers. Let's assume, completely arbitrarily, that the rock dates from the Jurassic period[31].

Finally, we assume that the entire African continental plate has acted more or less as a "rigid plate" since the formation of our now sampled rock. Such an idea greatly simplifies things. Especially in the border areas, the African plate will certainly have changed, for example by "growing" new oceanic crust, as discussed above, or by deformation or even subduction in areas of collision with neighboring plates. We also want to ignore such changes in the border areas in order to get to the point: the application of paleomagnetic information in plate tectonic reconstructions.

The sampling is done in exactly the same way as we have already summarized in the discussion of magnetostratigraphy. The samples are measured before extraction and provided with an orientation arrow, brought to the paleomagnetic laboratory, subjected to demagnetization experiments, and finally the magnetic vector is determined, which we equate with the orientation of the Earth's magnetic field at the time of the rock's formation. Of course, we have ensured, to the best of our geoscientific knowledge and conscience, that there is no doubt about the orientation of the determined magnetic direction.

This magnetic vector is known to have a declination and an inclination (Fig. 3.11). It is incidentally the exact same magnetic vector that we would have determined from this rock in a purely magnetostratigraphic investigation. However, we now consider this "arrow" from a different aspect. It is not interesting for this question whether the magnetic direction arrow reflects a configuration of the Earth's magnetic field that corresponds to the current arrangement of the North and South poles. However, it is very interesting what inclination angle we have determined based on the samples.

Let's assume we have taken our samples at a location at 30° north latitude (Fig. 4.19)— this could be, for example, roughly in the middle of Algeria. We now calculate what inclination angle we can expect (for a temporally averaged) Earth's magnetic field (GAD; Chap. 2) (that would be around 16°, but the exact value is not of further interest).

Now there are two possibilities.

The inclination value determined from the samples corresponds to the calculated value. There is then no reason to assume that the sample location has changed its geographical latitude since the formation of the sampled rock. If the sampled rocks, for example, come from the Tithonium stage of the Upper Jurassic, no significant movement of the

---

[31] For simplicity's sake, we ignore that the "Jurassic period" spans a period of about 55 million years, and assume that this is not particularly long in geological terms.

Fig. 4.19 The shift of a location on an assumed rigid Earth plate shifts this entire plate, depending on the result of the magnetic measurement, parallel to the longitudes into more northern or southern regions. For clarification, the relative change of the location at which the magnetic data were determined is also shown in dashed lines next to the outlines of a continent. According to the discussion in the text, the example of a location at around 30° north latitude on the African continent is shown. Thin black arrows indicate the relative direction of movement

entire (!) African continent would be expected for about 150 million years. After all, we consider Africa, according to our assumption, as a rigid geological unit. If the location of a place does not change, the relative position of any other place on the rigid plate also remains unchanged. No change in latitude, to be precise. But more on that shortly.

If the determined inclination deviates from the calculated inclination, the latitude of the sample location must have changed in the last—as in our arbitrarily chosen example—150 million years. If the determined inclination is flatter than the expected value (in our example 16°), the sample location in the Upper Jurassic would have been further south than today. If an inclination steeper than in our example—16°—were to result, a more northern location of the center of Algeria 150 million years ago would be inferred. The exact latitude of the shift to the north or south can again be calculated from the determined inclination value—with the same mathematical formula (not further specified formula), with which we initially calculated the expected inclination for the sample location. Only the other way around, because this time the unknown in the formula is the latitude and not the inclination.

4.2 Moving Continents with Stored Earth Magnetic Fields

Not bad, to demand the shift of an entire continent based on a 2 × 2 cm rock sample, right? We have now understood the principle of plate tectonic application. As always, however, we need to take a closer look at this point in order to do justice to the reality of geological research and its application, even in the slightest.

Let's stick with the idea of a rigid continental plate, exemplified by the dimensions of Africa (Fig. 4.19). Following our example, we now know the position of the continent about 150 million years ago based on the magnetic measurements. But how do we know that the continent has moved in a straight line to its current position, much like our above thought experiment about the supposed intermediate stations of a journey, as in Fig. 4.14. If we only have these two pieces of information—position today and position in the Upper Jurassic—the actual path between these two locations remains indeed a mystery, and one would have to be satisfied with the justification that in case of doubt the simplest solution is to be assumed, as there is no reason to assume anything else based on the data situation. The simplest solution is the direct path along a longitude in a northern or southern direction, as mentioned.

At this point, we can address a supposed problem with the correlation between magnetic inclination and geographical latitude. The calculation of latitude from the angle of magnetic inclination will always yield the correct value—however, the correlation is not suitable for determining at which geographical longitude the result is accurate. Fig. 4.21 illustrates why it is essential to consider the geological context when interpreting magnetic results. This is possible, for example, when a path of plate tectonic movements can actually be determined for a piece of crust, i.e., there is a sufficient number of sample locations and rock ages to trace plate tectonic movements (see also Fig. 4.22).

A path determined in this way can be checked for plausibility, for example, with a classic geological approach, by comparing the rocks of a locality with the supposed latitude at which these rocks were formed. In the context of further geological facts, for example, the deposition of typical glacial moraines at equatorial latitude would be unlikely.

A rather mathematical test of geological plausibility is to calculate the average speed that would result from the duration between two different locations of the supposed plate movement. Especially if inappropriately high speeds would result for movements of the Earth's crust for a subarea of the path, caution is certainly advised. Often, however, it will still not be possible to determine unequivocally based on the determined speeds whether there have not been "smaller" zigzag movements of a continent. This would be the case because a still plausible, for example, twice as high speed of plate movement, as would result for the direct line, also allows a twice as long distance to be covered.[32]

---

[32] As if I ride my bike leisurely in 10 minutes in a straight line from A to B, but with a little more effort and correspondingly higher speed, I can take the detour through the park in these 10 minutes before I arrive in B. If the required speed to manage the detour through the park, however, lies above realistic bicycle speeds (e-bikes included or excluded, we understand the principle), the path must have been more direct.

Doubts about the accuracy of the statement can only be dispelled for continental-scale tectonic plates if we have the opportunity to examine additional magnetic samples, specifically samples from rocks with ages between today and the Upper Jurassic (to stay with our example), so to speak, to insert smaller intermediate steps between the two endpoints (today, Upper Jurassic). Since we continue to consider the continent as a rigid unit in our example, we can search the entire continent for suitable rock layers. The same magnetic criteria apply to each new sample location as already described for the first location in our thought experiment. As with the first sample location, each magnetic result ultimately applies to the entire continent (Fig. 4.19). If all these samples each indicate movement along a direct line between the current position and the position in the Upper Jurassic, we have a better basis for the original assumption of movement along a longitude. Such a direct line would be indicated by continuously increasing or decreasing angles of inclination, as the location would have moved to continuously higher or lower latitudes, with the increasing age of the rocks.

If the magnetic results of the individual rock layers do not contradict a direct line, we simply have to move the continent along the corresponding position points, like along a string of pearls between today and the Upper Jurassic. Such a result would accordingly be a mixture of flatter and steeper inclination angles. So far, we have been moving the continent in our minds parallel to the longitude on which the sample location was at the time of sampling.

To cover a longer distance, i.e., to reach a neighboring longitude, there are two possibilities of movement:

The continent moves along the latitude and then again in a northern or southern direction parallel to the new longitude. Such a stair-like movement, similar to the translational movements in early computer games, is not in the geological nature of plate movements on a spherical surface.

The entire continent performs smaller or larger rotational movements on its way (as if a car were turning in a different direction). Such types of movement of parts of the Earth's crust are geologically more likely. The pivot point of these movements does not necessarily have to be in the center of the continent. Fig. 4.20 explains that the position of the pivot point in its location relative to the sample location must be taken into account, as only a sample location directly on the pivot point will not change its latitude solely due to the rotation (see also Fig. 4.23)[33].

---

[33] On a spherical surface like the Earth's surface, movements of continents occur on a large scale around a so-called Euler pole and thus are rather circular movements. An Euler pole may be located outside the continent under consideration, as if one were moving an object on a string around the nailed end point of this string. In this way, every movement of the continents on the Earth's surface is always a combination of translation and rotation and must be taken into account accordingly in plate tectonic models. Whether the proportion of translation or rotation represents the more decisive component is to be decided in each individual plate tectonic case. For a better understanding of this statement, imagine the movement of an object (or even a horse on a long

Fig. 4.20 Impact of a rotational movement on an Earth plate. A denotes the pivot point of the movement. For a better understanding, take a piece of cardboard and fix it with a nail on a base (remember, the base will be perforated). The nail represents the pivot point A, no matter where on the cardboard we have driven the nail. Point A retains its latitude even during a rotational movement. All other points on an Earth plate (or the cardboard)—here exemplified with B and C—will change their latitude due to a rotational movement. How this happens depends on the position of the pivot point A, the shape of the Earth plate (in other words, the position of the considered points B, C relative to the pivot point A) and to some extent the initial position of the Earth plate. The larger an Earth plate is, the more dramatic the effects will be, but the principle applies in the same way even to very small fragments of the Earth's crust, which move independently. Most movements of the Earth's crust will represent a combination of a movement along a longitude (Fig. 4.19) and a rotational movement (see also Fig. 4.21 and 4.22). All relative combinations of translation to rotation are possible, which is explained in more detail in the text under the keyword "Euler pole" (see also Fig. 4.23)

This mild criticism of our own assumption that we can easily move entire crustal areas the size of the African Plate based on a single sample point, however, leads us to another point in considering the geological reality in the interpretation of our paleomagnetic results.

---

leash; Fig. 4.23) and what effect it has when you change the length of the string (again one of our somewhat far-fetched analogies in this book). On a very long string (leash), the object (or the horse) moves on a circular path with a huge radius and thus, at a momentary consideration, seemingly hardly noticeably on a circular arc. On a very short leash, the horse moves on a very tight circular path around us, with a relatively high proportion of "rotation" to get from one point to another.

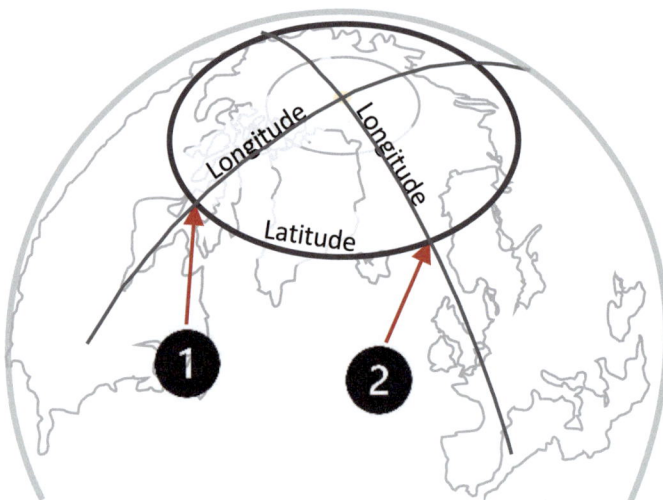

Fig. 4.21 The correlation between magnetic inclination and geographical latitude discussed in the text (see also Fig. 2.7) leaves open in the result which intersection (exemplarily here ①, ②) with a longitude on the calculated latitude circle is to be assumed for the position of a plate. However, each tectonic plate moves slowly and along a continuous path, and therefore it is not geologically or correspondingly plate tectonically sensible to assume that a continent has "hopped" randomly to any positions along a latitude during its journey. For plate tectonic models, it is also mandatory to consider the movements of all known tectonic plates in context. Strictly speaking, the movement of a plate always has a consequence for the position and movement of all other plates on Earth, as long as the movement cannot be compensated by deformation or subduction in the border areas of the neighboring plates. The plate tectonic interpretation possibilities of magnetic results are thus ideally restricted in a favorable way

As already mentioned, all areas of the Earth's crust resemble a geological patchwork. In other words, much smaller sub-areas than the African Plate must be examined to account for the geological complexity of the development of the Earth's crust over the course of Earth's history. Only some areas of the Earth's crust and thus of the continental plates are made up of rocks billions of years old. Individual areas of continental or oceanic crust have only emerged later and at different times in Earth's history or have been subducted at some point in Earth's history and are thus accessible to paleomagnetic investigation or plate tectonic reconstruction. Therefore, the magnetic investigation of rocks will aim at as many geological rock formations of different ages as possible with a worldwide distribution. Only in this way can a broad and verifiable database be created. In combination with all geological and geophysical data of a region, this also allows to narrow down periods in which individual crustal areas followed a common path, only to "pursue" different paths from a certain point in time. Typical examples of such a situation arise when investigating the breaking apart of larger continental plates, like the massive landmass Pangea in the Permian. In the northern border areas of the African Plate,

4.2 Moving Continents with Stored Earth Magnetic Fields   135

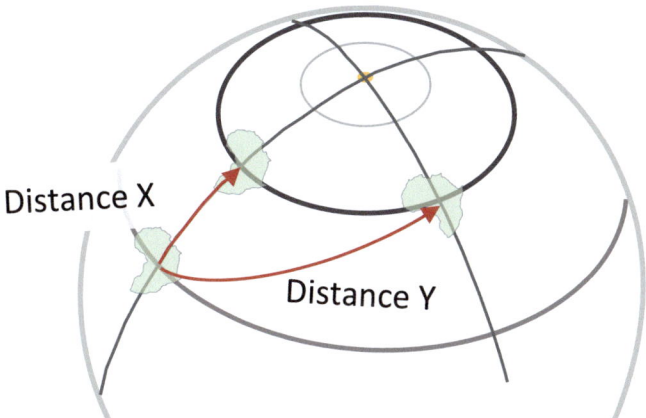

Fig. 4.22 A rather mathematical way to restrict the possible positions calculated from the inclination on a circle of latitude is to determine the maximum distance for the movement of a continent in a certain period of time (as in our thought experiment in the text, for example, 150 million years). To do this, one determines the direct movement along a longitude circle as the shortest distance X and the maximum possible distance Y based on geologically indicated movement speeds of the Earth's crust. Such speeds are, as mentioned, on the order of centimeters per year. Therefore, the method works better the shorter the period considered. Even a speed of centimeters per year is enough, over the duration of Earth's history, to let a continent "circumnavigate" the Earth several times

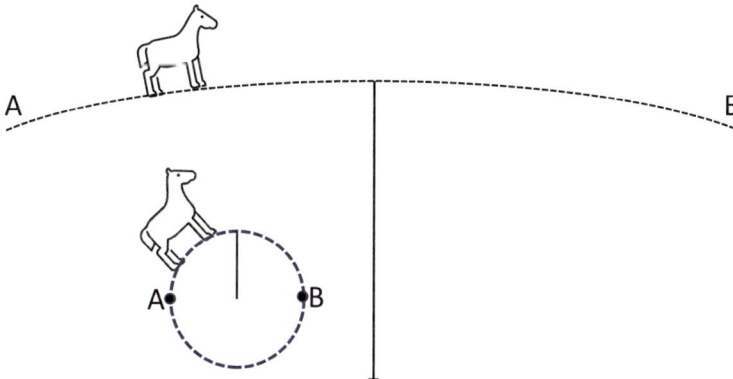

Fig. 4.23 As a (somewhat limping) analogy to the relative importance of translation and rotation in the movement of continental plates, the figure shows the path of a horse on a circular path with a very small radius compared to a very large radius. In the latter case, the horse seems to run, without leaning, on an almost "straight line" from A to B. When interpreting plate tectonic movements from records of the Earth's magnetic field, the effect of rotation is always important when a crustal fragment has not moved strictly parallel to a longitude or a latitude

smaller pieces of crust "broke off" so to speak[34], to then make their way independently of Africa's later movement within the framework of global plate tectonic movements. There are much smaller geological units than the terrains implied here[35] Avalonia or Armorika, which later became parts of other land masses through collision and then, sometimes heavily tectonically deformed, followed the path of one of these land masses for a time.

In the extreme case, even the interpretation of the geological regional environment does not allow a clear classification of the positions of a piece of the Earth's crust. In other words, even if a latitude for a crust fragment can be calculated from the magnetic inclination at a certain point in Earth's history, its exact position remains quite vague. Such cases will occur more frequently the smaller and older these examined, supposedly independent pieces of crust are. In any case, based on a scientifically and conscientiously determined magnetic inclination value, the latitude of a crust fragment can be determined at a certain point in Earth's history. In the case of an otherwise sparse geological data situation regarding this crust fragment, this may be one of the few concrete pieces of information available to integrate this fragment into the overall plate tectonic picture. Information that is based on a concrete inclination value and whose significance is therefore scientifically robust.

An important point in the plate tectonic reconstruction of Earth's history using magnetic direction data needs to be emphasized again. So far, we have assumed that the magnetic value determined from a rock layer corresponds to a temporally averaged geocentric axial dipole field and the corresponding magnetic pole thus corresponds to the geographical North Pole.

We already suspect that upon closer inspection this is a strong simplification. After all, we have thus assumed that in a typical (2 cm large) magnetic sample the data of the Earth's magnetic field from several thousand years are stored (Chap. 2), in other words, that the formation of these 2 cm of rock took many thousands of years, so we determine the average direction of the Earth's magnetic field from this period in a magnetic measurement.

Let's imagine as a rather unconventional analogy in everyday life a piece of cake consisting of different layers of dough, chocolate and strawberries: Eating all layers at once corresponds to measuring the averaged GAD field. If we only pick out the strawberries from one layer, we "taste" only the strawberries, similar to when we measure a smaller, magnetic sample and then determine the recording of a shorter period of the Earth's

---

[34] "Rifting" is the more geologically precise term.

[35] Terrains, sometimes also called terranes, are crustal areas which have followed independent paths relative to the main plates in a plate tectonic context—micro-earth plates, so to speak. There are many more of these than the seven main plates of the Earth's crust mentioned.

4.2 Moving Continents with Stored Earth Magnetic Fields

magnetic field for the same rock—a period that may be too short to meet our criteria for a GAD field (Chap. 2).

Unlike magnetic samples, with pieces of cake you have the choice of how much to bite off. In geology, an optimal path must be found from the ideally measured sample size, different durations of formation of different types of rock, and many other criteria. Consequently, this also means that in many cases of magnetic sampling it will not be possible to meet our assumption of a GAD field—simply because no rocks from a sought-after period with a sufficiently low accumulation rate are exposed on the examined crust fragment. For the application to plate tectonic reconstructions, this means two seemingly disadvantageous things:

We discussed earlier that the current Earth's magnetic field does not completely correspond to a symmetrical dipole (GAD) and therefore cannot be correlated directly with the geographical coordinate system. If I go outside today, wherever I am, sample "recently" formed rock and determine a magnetic inclination angle from it, a calculated magnetic pole will not coincide with the Earth's geographical pole (as in Fig. 2.7). This pole is somewhere off quite a bit (as in Fig. 2.3). In paleomagnetism, such a pole is also referred to as a virtual geomagnetic pole (VGP).

A latitude position of a crustal piece determined in this way from the inclination value does not match the latitude positions of the GAD model for a certain period of time. The VGP does not coincide with the axis of rotation (Fig. 4.24).

So far our little insight into what the storage of past states of the Earth's magnetic field in rocks can do for plate tectonic reconstructions. In summary, the following points should be emphasized:

- The magnetic "arrow" stored in rocks can show us the geographical latitude of a rock at the time of its formation in Earth's history.
- Depending on different types of rocks, an Earth's magnetic field averaged over longer periods of time can be stored, which can be equated with the field of a geocentric axial dipole field (GAD).
- The symmetry of the GAD allows a correlation between geographical latitude and the magnetic inclination at this location to be established for every point on Earth.
- If the rock was formed "too quickly" to get a sufficiently long view of the highly variable Earth's magnetic field in Earth's history, a "virtual" latitude position is determined from the magnetic inclination.[36]
- The determination of the "magnetic arrow" or magnetic vector as a representative of the magnetic field direction must be determined through careful laboratory experiments (demagnetization experiments!). The sought-after information is the magnetic declination and inclination at the time of the rock's formation.

---

[36] In other words, a latitude calculated purely from the measured values probably has no relation to the actual place of origin of the rock.

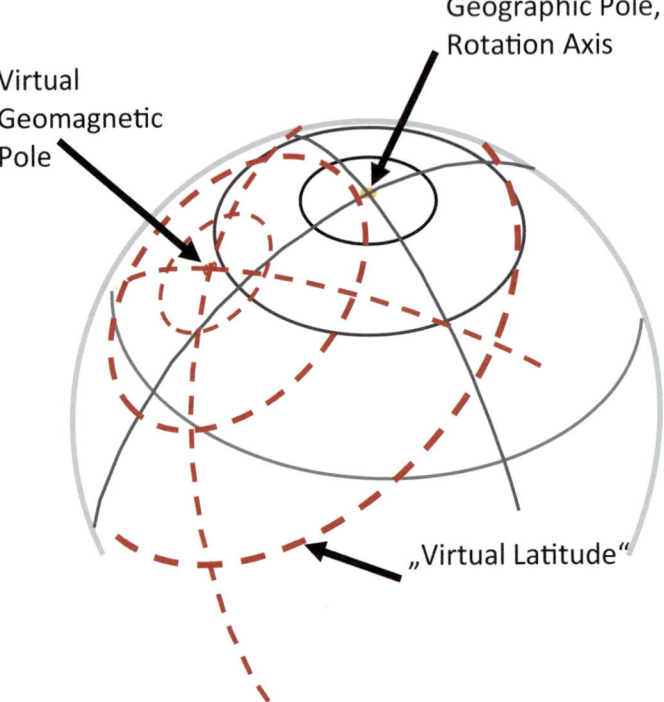

Fig. 4.24 Only when the determined "magnetic arrow" also represents an averaged recording of the geomagnetic field in the sense of a geocentric axial dipole field (GAD), does the latitude position calculated from the inclination correspond to the sought-after geographical latitude position. In all other cases, a virtual geomagnetic pole (VGP) is calculated and thus a virtual latitude position (dashed lines) is given for the examined crustal fragment. However, note that a multitude of independent crustal fragments from rocks of the same age on average yield a geographical (GAD-) pole again (see Fig. 2.8). With the knowledge of having determined only a virtual latitude position due to the rock properties for a crustal fragment, the search for further rocks of the same age but with as worldwide distribution as possible at the time of their formation must be carried out

- Deviations of the declination from the north direction prove a rotation of the examined crustal piece in relation to the geographical system of longitude and latitude.
- Any interpretation of determined magnetic values and their conversion into geographical positions must be in strict unity with all other available geological information about the rocks. In particular, the supra-regional plate tectonic context is of importance. This is not a particular weakness of paleomagnetic data, but a direct consequence of the complex geological history of our planet.[37]

---

[37] In other words: Always apply a sanity check (see above) in geoscientific studies.

4.2 Moving Continents with Stored Earth Magnetic Fields

- Successful plate tectonic models are based on the reconstruction of a multitude of smallest crustal pieces (and these are significantly more than the seven main tectonic plates) over the entire (as long as possible) period of Earth's history. Such models are used, for example, to explain the position and formation of mineral deposits or to highlight earthquake risks in the plate tectonic context.

To conclude this chapter on the use of stored directions of the Earth's magnetic field in plate tectonics, a brief note: We have highlighted the enormous importance of the geological context in the interpretation of magnetic data. In a directly apparent form, taking into account the geological context also means considering the tectonic displacement of a rock layer. Here again, it helps to imagine the stored information of the Earth's magnetic field as an (imaginary) arrow in a rock layer.

As a rule, rocks are horizontal at their formation, and in this orientation of the layers, the magnetic vector is also recorded (e.g., Fig. 2.20). If the relevant rock layer is tilted at a later point, for example during crustal extension, the magnetic arrow is also twisted. It is geological standard to measure such layer inclinations with a suitable measuring device (geologist's compass). Accordingly, a specific angle can be given for each rock layer, under which the layer dips from the horizontal. Similarly, a second angle is determined, which indicates whether the dipping of the layer simply occurs to the north or in any other direction. Before a magnetic arrow can be interpreted in its orientation, any tectonic layer adjustment must be reversed (Fig. 4.25—as always, this means on paper).[38]

In conclusion, two essential points of geological, detective work in this type of paleomagnetic studies can be summarized as follows:

- Verify the temporal correspondence of determined magnetic direction and rock age— understand which phases of tectonic stress could have led to mineralogical changes due to increased temperatures or pressures. In everyday life, such a review of the backgrounds of any kind of "appearance and reality" corresponds.

---

[38] Specifically, this means that the inclination and declination of the magnetic vector are corrected by the measured angles of layer inclination. We do not need to explain this procedure in detail here; ultimately, it is just a mathematical step called coordinate transformation. Angles in three-dimensional space, as in the comparison of magnetic inclination and layer inclination, cannot simply be subtracted one to one from each other. However, if the order of the necessary steps is observed, a correction of the inclination values is routine work and can be easily calculated. Sometimes, several consecutive coordinate transformations are necessary to obtain the correct value of the inclination. Many rocks, for example, have been displaced in several consecutive deformation phases over the many millions of years since their formation and storage of the Earth's magnetic field. Correcting only the last phase does not necessarily return layers in such cases to their original horizontal position.

140   4 Paleomagnetism—Applications

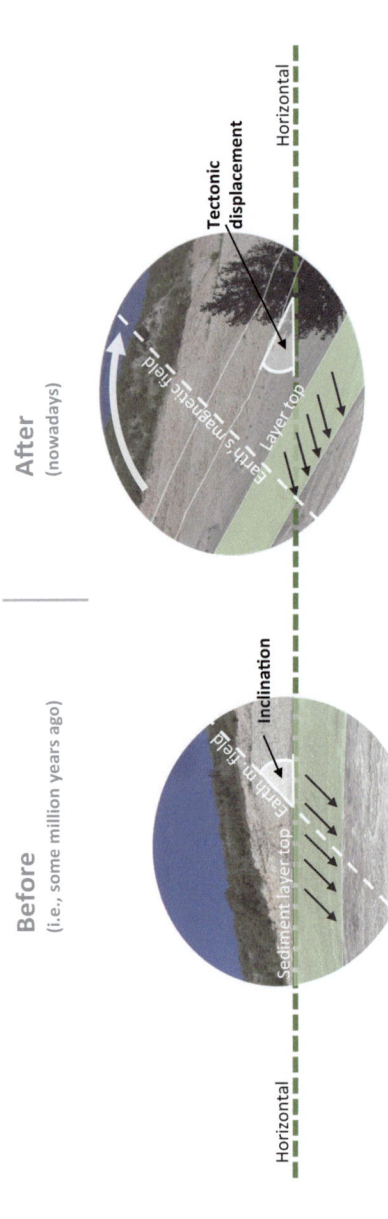

Fig. 4.25 Illustration of the tectonic correction of a stored direction of the Earth's magnetic field. At its formation, a rock layer acquires a magnetization in the direction of the Earth's magnetic field (black arrows) at this location (before). The inclination of the magnetic direction is the angle between the field line of the Earth's magnetic field and the horizontal. The angle of incidence of the magnetic field line is linked to the geographical latitude of the location. Our schematic drawing could be located at about 30° northern latitude based on the fictitious field lines. **After** the formation of the layer, further rock layers were deposited, but, in particular, the layers were tilted in the course of tectonic events (after). The stored direction of the Earth's magnetic field in the old layer was thus also displaced. It no longer corresponds to the direction of the Earth's magnetic field at this location. The angle of tectonic displacement is the key to correcting the magnetic results

4.2 Moving Continents with Stored Earth Magnetic Fields

- Determine the originally horizontal position of the rock and thus of the magnetic vector—understand the tectonic phases and the geometric effects on the storage of the rock. If you confuse left and right in everyday life when navigating, you will reach a destination, probably not the originally sought one though.

> **Direction and Reference System**
>
> Direction implies a reference to a reference system in which this direction is defined. In the case of the Earth, a possible such reference system might be the geographical system with north, south, east, and west plus up/down, i.e., the reference to longitude and latitude. For example—since there are alternative ways to subdivide the Earth's surface into smaller areas—to define a position on Earth with an indication of the type "so much in one direction and then so much in the other direction" (examples are the UTM grid or the reference to longitude and latitude). All reference systems have the disadvantage of not showing the same precision at every point on Earth, even if they do not move relative to each other like the tectonic plates. For understanding: The distance from one longitude to the next is about 111 km at the equator. At the geographical pole, one can hop from one longitude to the next. In the context of this book, it should only be noted that it is by no means trivial to define fixed reference points for the Earth's body as a reference system for all movements on Earth for geological periods. As already noted, the Earth, including the Earth's crust with its continents and ocean floors, is a very dynamic system in constant change, whose components are continuously moving around each other.
>
> In any case, two or more directions can only be compared with each other if they refer to the same reference system, and that also in terms of time, for a point in geological past. Imagine a reference system chosen from the three places New York, Cologne, and Marrakech. A position indication in relation to these three places (triangulation; Fig. 4.26) would not be very accurate in a few years, as both the continental plates of North America and Europe and these in relation to Africa are moving (tectonically, geologically speaking) at speeds of a few centimeters per year relative to each other. What would such a position indication for a point 100 million years ago, in the Cretaceous period, be worth? Not very much obviously, because in this period the three corner points would have potentially moved several hundred kilometers relative to each other (Fig. 4.26). Apart from the fact that neither New York nor Marrakech existed in the Cretaceous period, but that's not the point.

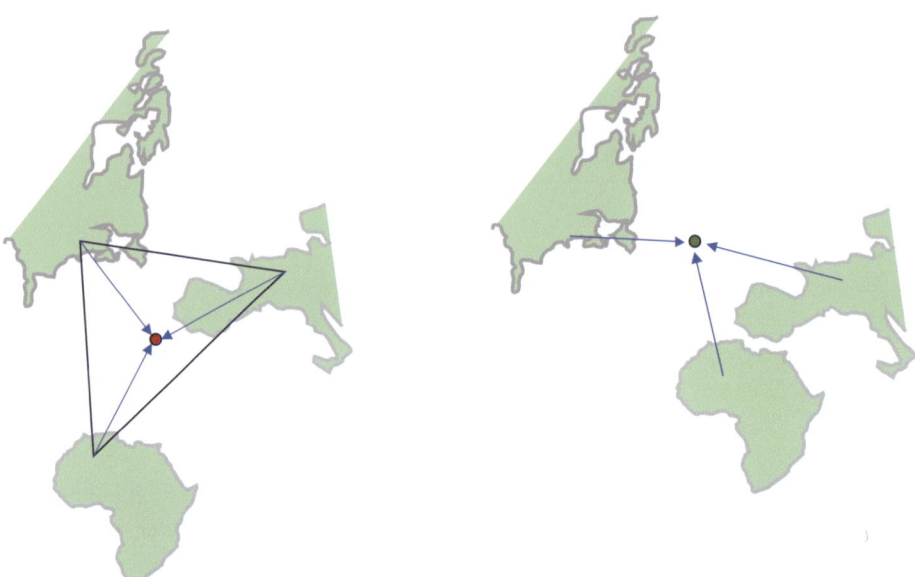

Fig. 4.26 Determining the position of a point by triangulation is affected on a geological scale by the relative movement of the tectonic plates. A point in relation to America, Africa, and Europe (left) is hard to find millions of years later (right)

> **Declination and Inclination in a Pole Reversal**
>
> A compass needle only swings horizontally at the equator. In Fig. 4.2 this is represented by an arrow parallel to the Earth's surface. Towards the pole, the arrow takes an increasingly steep angle. In other words: the compass needle aligns itself along the field lines of the Earth's magnetic field, not only by pointing north (south), but also by taking the inclination angle corresponding to the latitude. In paleomagnetism, this inclination or dip angle towards the Earth's surface is called *inclination*. Especially for a temporally averaged and thus symmetrical dipole field, as we will often find it stored in rocks, some very useful rules can be derived from this.
>
> In the case of a pole reversal, the inclination would remain (considered as an absolute value), but our imagined arrow would no longer point to one pole (let's say north) but to the opposite pole (let's say south).
>
> The *declination* thus changes by 180°, as the magnetic needle, and thus the arrow will now point to the opposite pole. Accordingly, imagine all the arrows in Fig. 4.1 in the opposite direction. Capturing such "declination swings" over the course of Earth's history is the basis of a dating method established in the geosciences—magnetostratigraphy.

Just as the field lines dip almost vertically at the pole, a freely swinging compass needle would accordingly settle in. If a rock acquires its magnetization at the pole, the acquired magnetic inclination of the rock will also be 90° to the Earth's surface. If the same rock were formed in Cologne, the stored angle of the rock's inclination would also be flatter (about 30°) due to the lower latitude (Cologne is at about 51° latitude). Latitude and magnetic inclination are not exactly the same, but they are directly related and are therefore easy to calculate from each other. Regardless of the dip angle of the inclination, all magnetic needles at any position on Earth align parallel to the Earth's longitudes. As we know, the compass needle shows us where north is. Accordingly, the magnetization acquired in a rock will point towards the pole at any location in the world. The arrow points instead of north, now towards the South Pole, and vice versa. Rocks that already had acquired a magnetization will not be influenced per se by a polarity change. However, newly formed rocks will record this new changed direction of the Earth's magnetic field. The importance of reference systems in magnetism can now also be guessed. For your own considerations: How does the magnetization of a rock, which was acquired at the same time on the corresponding latitude north or south of the equator, differ?

Rock Magnetism 5

5.1 Delineation from Paleomagnetism

In Chap. 4 we proposed to approach the term "paleomagnetism" in a classical etymological way first. A similar approach to explain the term "rock magnetism" may seem less suitable as an introduction to this chapter. Rock and magnetism—both words, individually or combined, seem to describe exactly the topic we have extensively dealt with in Chap. 3. We have also already received a somewhat casually formulated paraphrase of the term "rock magnetism" there: "How do rocks do that with magnetization?"

We do not want to get bogged down with definitions and delineations within a geoscientific discipline, so just this much here in the sense of this book and as an introduction to this chapter: Paleo- and rock magnetism are often mentioned in the same breath in scientific language when it comes to describing the topic of rocks of the Earth's crust and their magnetization. Both disciplines are closely linked through the findings from the research of the physical magnetic properties of rocks. Many sections in Chap. 3, which deal with the magnetic properties of rocks, the different acquisition mechanisms, the magnetic properties of specific minerals, the effect of grain size on magnetic properties, etc., belong with equal justification in Chap. 4 and in this chapter on rock magnetism.

In this book, we use the two umbrella terms "paleomagnetism" and "rock magnetism" in a very pragmatic way as follows:

- **Paleomagnetism** we use when we investigate the direction of the Earth's magnetic field stored in rocks from the Earth's past. In Chap. 3 we used the rather math-heavy term "vectors" (Fig. 3.14) for this. Therefore, paleomagnetic studies are based exclusively on a very small—albeit geologically influential—group of minerals that can store the Earth's magnetic field

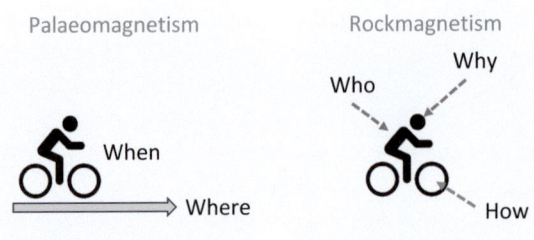

Fig. 5.1 To distinguish between paleomagnetism (Chap. 3; left) and rock magnetism (this chapter; right) in a bicycle analogy. Neither can do without the other in science or in sports

- **Rock magnetism**, in contrast, is intended to describe all such studies and geoscientific applications that deal with the magnetic properties of all minerals, especially when these minerals are incapable of a permanent recording of a magnetic field and thus a direction. Compared to the term "vector", such properties can be referred to as scalars. To stay with the image from Fig. 3.14, it is not of interest in which direction the bicycle is steering, but for example, how fast it could go or how aerodynamic the bike is. Summarized and translated, the question is primarily about what kind of bicycle it is (Fig. 5.1).

A translation of this renewed bicycle analogy into geological and indeed rock magnetic questions is easily done: We want to determine the composition of rock in the geological layers being studied through magnetic measurements. No, not in the sense of naming a rock as basalt, granite or sandstone—there are numerous better methods for that. In case of doubt, a person with geological training, a hammer and a magnifying glass are sufficient.

By composition of rocks, we mean rather to determine whether the composition[1] of a similar type of rock changes relative to each other from layer to layer. In the previous chapters, we have already addressed some geological scenarios to highlight that changes and variations in the composition of rocks always underlie changes in the environmental and deposition conditions of these rocks. Some such differences in the rocks are primarily present since the formation of the rock (think for example of variations in the delivery of sediment components) or perhaps only later caused by weathering or contact of the components with fluids in the subsurface (think for example of groundwater flows).

---

[1] As a reminder: Most rocks—magmatic, sedimentary or metamorphic, regardless of their provenance and texture—essentially consist of individual mineral grains of different types.

To appreciate the last statement, we first need to return to a strongly geological, at least geoscientific, perspective and highlight two essential points:

1. Each type of rock can be assigned to typical geological scenarios and thus also (past) environmental processes. Basalts obviously form in a different geological environment than sandstones. But even within a similar depositional environment and similar types of rock, the smallest changes in environmental conditions affect the composition (the mineral content) of the rock. Such changes in a deposition space provide information about a temporal and spatial part of the earth's historical development.
2. Each mineral has a typical magnetic fingerprint. This refers to mineral-specific magnetic characteristics (Fig. 4.2). The fingerprint results from a combination of mineral type, chemical and physical deviations from the ideal growth of this mineral, and the size of a mineral grain (Chap. 3). Even minor changes will noticeably alter a fingerprint. In particular, remanent magnetic minerals in rocks are particularly susceptible to changes—or even dissolution—due to their small size in the nanometer range, but also, for example, their composition. For instance, such minerals usually contain iron in their crystal lattice, and indeed, rocks do rust in a certain way. Or they are eaten (into) by the most diverse types of bacteria (yes, indeed).[2]

Ideally, we will use the rock magnetic properties of rocks to shed more light on the geological development of a part of the Earth's crust. Apart from explaining the rock magnetic properties of minerals and rocks, particularly geoscientific applications are, against this background, referred to as environmental magnetism.[3]

At first glance, it may seem more immediately understandable why we explained the storage of the Earth's magnetic field in rocks in Chap. 2, 3 and 4. After all, the Earth's magnetic field provides a unique signal in the rocks and thus information about past states of our Earth that is not accessible in any other way. Isn't it therefore "ambitious" to want to characterize the composition of rocks through magnetic measurements, when there are numerous and obvious other geoscientific methods to do this—mineralogy, microscopy, geochemistry, to name just a few?

Two reasons can be quickly mentioned:

- Magnetic methods are easy to apply and non-destructive for the sampled material. Other methods are potentially also; our rock magnetic approach would therefore have to be at least simpler, more cost-effective, faster or ideally—provide insights into the world of geosciences that cannot be achieved in any other way.

---

[2] Only the small magnetic particles, not the entire rock. To preempt any panic. And only as long as water can still circulate in the pore spaces of the rock. More on this in Sect. 5.2.

[3] For example, Evans, M.E. and Heller, F. 2003, *Environmental Magnetism: Principles and Applications of Enviromagnetics*, Academic Press, 299 pages.

Thompson, R. and Oldfield, F. 1986, *Environmental Magnetism*, Springer, 227 pages.

- Rock magnetic measurements are very sensitive and thus high-resolution in geological details (explanation follows), but so are other non-magnetic methods in geosciences. In one way or another, rock magnetic methods are indeed important in geological practice and are certainly welcome in terms of finding results. For a really compelling reason to undertake a rock magnetic characterization of a rock sequence, we therefore need to delve a little deeper into our considerations.

In Chap. 3 we classified the entirety of all minerals into three magnetic groups: dia-, para-, and ferromagnetics (Fig. 3.2). Only a small number of minerals from the group of ferromagnetics can record the Earth's magnetic field and preserve it over geological periods. Direction-independent magnetic parameters—in Chap. 3 we introduced magnetic susceptibility as a scalar measure of the magnetizability of a substance—are however influenced by all minerals in a rock, thus the dia-, para-, and ferromagnetics. For example, the quartz grains of a sandstone also contribute to the magnetic susceptibility, while the same quartz grains have no part in recording the Earth's magnetic field in this rock (Chap. 2, 3 and 4). Quartz is diamagnetic and certainly not a remanent magnetic mineral. Nevertheless, the magnetic susceptibility of a sandstone consisting of over 99% quartz grains is not necessarily controlled by these main components of the rock. The answer to this apparent contradiction lies in the different—greatly different—intensity of the magnetic susceptibility compared to dia-, para-, or ferromagnetic minerals. In Sect. 3.5 we considered in more detail how the contributions of different minerals affect the measurement of magnetic susceptibility.

At this point, we assert—as always, plakatively, while we stay with an ordinary sandstone as an example despite the multitude of different rocks of the Earth's crust: Even the smallest proportions of remanent magnetic minerals control the results of all types of magnetic measurements on rocks. And only for this reason are many rock magnetic applications in geology possible at all. What does this mean again?

The measurement of the magnetic susceptibility of a rock provides a single numerical value in the simplest case. In most cases, however, this numerical value should and can not provide a direct result in the sense of a rock identification. For example, it is not possible to measure a sample, obtain a value X^4 and to conclude that sandstone, granite, or any other rock is present. Even if the results for sandstones and basalts do sit in different

---

[4] Of course, in the correct physical unit of magnetic susceptibility. The unit is not of concern for our basic understanding, as noted elsewhere. Bulk susceptibility in the so-called SI system of units is actually dimensionless. Often a mass-specific susceptibility is used. The physical unit then has the form of volume per weight, for example m^3/kg. Important for our understanding is only that the magnetic susceptibility depends, for example, on the amount of the material examined. A larger chunk of the same material has a correspondingly higher susceptibility (see Sect. 5.2 for a practical application of this property).

5.1 Delineation from Paleomagnetism

ranges of susceptibility values, there is no exact numerical value, not even a sufficiently characteristic range of values, with which this rock could be clearly identified. And that is not necessary. The value of such measurements lies elsewhere.

A single value of magnetic susceptibility is something like a black box. We initially do not know what this value is composed of, only that each individual component of a sample, albeit to varying degrees, has contributed to our general result. The repeated measurement of susceptibility along a rock wall—that is, from one layer to the next rock layer—will, however, provide a sequence of higher or lower values relative to each other (Fig. 3.17). If the measurements are made across a sequence of layers, this sequence of layers provides a typical pattern of magnetic values over a period of Earth's history, without it being necessary to know what the reason for the fluctuations in the values is[5]. It is enough to compare the (variation) patterns of two supposedly identical rock sequences at the fictional locations A and B to correlate these sequences (more on this below). In case of discrepancies in the patterns, one goes in search of the geological causes if necessary.

The correlation of layers at different locations is the be-all and end-all of geological studies—how else would it be possible to create a coherent three-dimensional image of the subsurface (the basics of geology)? Sect. 5.2 explains one type of such applications in detail using the example of Buddha statues (!).

In a continuation of the idea of using magnetic susceptibility as a proxy for lithological changes in the Earth's crust, efforts will be made to open the black box in order to more accurately qualify and quantify the individual contributing components. In this book, we generally use sedimentary rocks to explain geological facts (Chap. 2, 3 and 4), and from these examples it becomes immediately apparent that the composition of the components of a rock (type, size, shape) is directly controlled by the environment—source, transport and deposition space—of the sediment. Different rocks are formed under hot climates in an arid desert than at the bottom of a glacial lake.

If rock magnetic studies want to contribute to the deciphering of environmental conditions in Earth's history, there is no way around characterizing the individual components of a rock sample. It should also be noted that our explanations using the example of sediments can also be applied to magmatic and, to some extent, metamorphic rocks, and thus all rocks. (For reminder: We divide the rocks of the Earth's crust into magmatic, metamorphic or sedimentary). The examples of applications would then simply have to be chosen differently.

However, using a bulk susceptibility value alone, it is not possible to characterize the individual components of a rock sample—additional, so-called rock magnetic measurements must be carried out. After many years of research in the field of rock magnetism, there are a number of laboratory procedures aimed at identifying certain minerals based

---

[5] The simplest reason for such would be, as said, changes in the type or composition of the successive rocks.

on their special magnetic properties beyond magnetic susceptibility. Many such measurements specifically target the small group of remanent magnetic minerals in rocks. All types of minerals contribute to magnetic susceptibility, but the magnetic properties of rocks are usually dominated by this illustrious group of seemingly underrepresented ferromagnetics in terms of size and quantity (Sect. 3.1).

It should be emphasized that this small group of ferromagnetics can tell at least the same geological story as the majority of the other minerals of a rock (for example, quartz grains). However, this small group of minerals has the advantage of having captured this history more sensitively and therefore being able to "report in more detail". Why this is so will become clearer in the course of the chapter, but as already indicated, it is mainly due to the peculiarity that even the smallest changes in these minerals will also change their magnetic properties. Therefore, it is not only justified, but also opportune for reasons of sensitivity of the results and possibilities of interpretation, to focus our special attention on ferromagnetics. Examples will follow.

Opening the black box of bulk magnetic susceptibility therefore means reading out this one mineral or, in other words, this illustrious group from the multitude of all components in a rock.

And this is initially not a trivial task, considering the example of susceptibility, that all existing minerals react simultaneously to this (magnetic) measurement, as if one were to identify a single child from a choir singing with closed eyes. At least if not only the main voice is sought, but the quieter thirds and fifths are to be determined, different technical aids including filters for individual tone frequencies will probably be needed.

A dominant (magnetic) mineral in a rock sample may still be easy to identify. A more precise characterization of such a mineral, for example according to grain size or mixture with other accessory minerals, possibly with similar properties, requires more elaborate investigations and measurements. Sometimes a series of different measurements and measurement methods are necessary, the results of which must be integrated and analyzed with almost detective-like meticulousness. Sometimes the effort is great—in terms of a cost-benefit consideration, however, the effort should not be too great—to get to the most interesting details.

In Chap. 3 we learned that we need to zoom into the atomic to subatomic level of minerals to explain the magnetic properties. It is therefore all too easy to see that even minor changes will lead to a change in the magnetic properties of these minerals. Changes in the minerals can refer simply to their grain size, but also to the nature of the crystal lattice (mixed series).

At this point, we want to highlight the following fact: A large part of rock magnetic basic research is to subject each available mineral type of the earth's crust to a variety of different magnetic measurements and to compile the results into a (magnetic) profile for this mineral after careful analysis (Fig. 5.2). Corresponding laboratory experiments take place under environmental conditions, as we typically find them on the earth's surface. On the other hand, the minerals are subjected to further physical procedures to see which

5.1 Delineation from Paleomagnetism

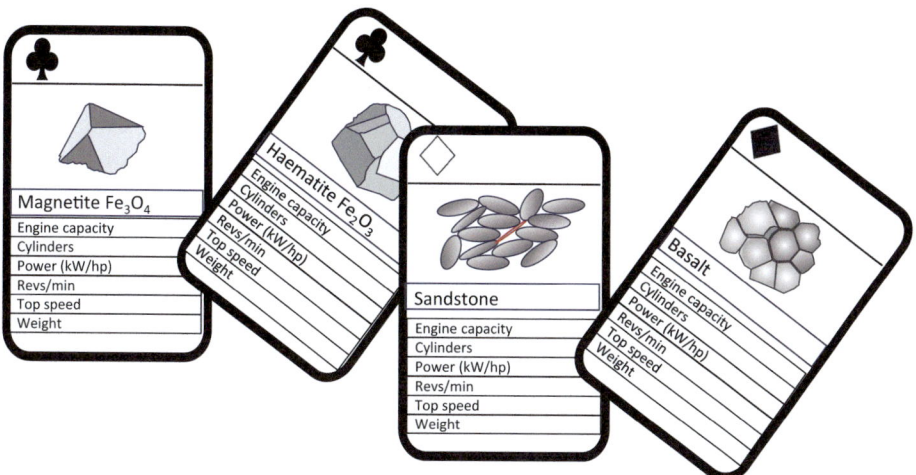

Fig. 5.2 Both rocks and magnetic minerals are characterized by a variety of individual properties. One goal of rock magnetic studies is to draw conclusions about geological processes through the knowledge, measurements, and analysis of corresponding characteristics

magnetic reactions occur, for example, when the environmental temperature or the surrounding pressure changes, etc.—all this, among other things, to determine whether a certain procedure or measurement leads to a unique and thus diagnostic magnetic characteristic for this mineral, as if only a certain member of the choir reaches the three-lined C and can therefore be recognized in the crowd.

It is very important for understanding such studies to remember that the objects of magnetic investigations are very small (magnetic minerals are often only 1/100 mm in size; Chap. 3). The amount and size of magnetite particles (or any other remanently magnetic mineral) present in a rock sample cannot be determined with a magnifying glass, not even with significantly higher-resolution electron microscopes. The best transmission microscopes make it possible to recognize even very small magnetite particles. However, the technical and time effort is then high, especially when a complexly composed rock sample is examined. It should also be considered that remanently magnetic minerals like magnetite are not only small in most rocks, but also very sparse (Chap. 3). To put it bluntly, one has to search for a long time to discover a magnetite particle among the other minerals of a rock sample—at least optically; a magnetic signal, on the other hand, reveals even the proverbial needle in a haystack. In practice, purely magnetic, in this sense deductive methods are often used to determine the composition of the magnetic mineralogy of a sample. And for this, we need diagnostically valuable rock magnetic characteristics of the minerals.

We now have numerous such characteristic magnetic characteristics of minerals available for "deduction". In many cases, such characteristics are based on so-called laboratory-induced magnetizations. This means that a sample is exposed to artificially generated magnetic fields. Such magnetic fields are usually much stronger than the earth's magnetic field when rocks are magnetized naturally. Since one can specifically control the direction, strength, and type of such magnetic fields in a laboratory, a variety of possible insights into the reaction of a sample to the experiments carried out arises. Often, the test parameters (field strength, direction, temperature, etc.) are changed during such experiments, either in discrete steps or continuously, as if the blood pressure is measured, an EKG (electrocardiogram) or a CT (computed tomography) is performed during an examination in a hospital.

Since diagnostic characteristics have been determined on an individual mineral, i.e., separated from the rock association, the challenge in geological and rock magnetic practice often is to filter out such information from the lot (the rock association). Therefore, one often additionally searches for typical characteristics in the geological rock world of mixtures of such characteristics. For example: What does the magnetic signal look like when a mixture of small (let's say, "superparamagnetic" particles; Chap. 3) and very large (multi-range particles; Chap. 3) grains are present in a rock? The results of similar magnetic studies on various rocks, compiled from all scientific publications in this field, can then lead to very informative, for example binary diagrams, to classify new results (Fig. 5.3).

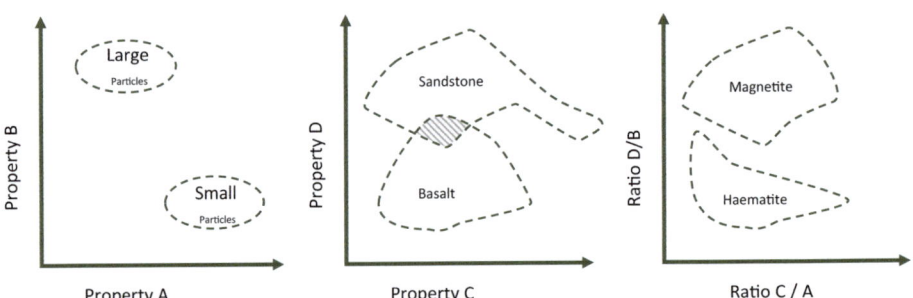

Fig. 5.3 Three examples of fictitious binary diagrams in rock magnetism. From special laboratory experiments or a review of scientific results of examined rocks, "empirical ranges" are marked, which show us, for example, that small single-domain particles of magnetite show high values of a property A, but low values of a property B (left). Other magnetic properties (C, D) are typically found in one type of rock, but not in others—however, this separation is not clear in the middle range of properties (center). Often, ratios are formed from individual properties to achieve—here exemplarily—a distinction of different remanent magnetic minerals (right). Such diagrams are one of the possibilities to meet the challenges in the deductive determination of a complex magnetomineralogy of a rock. Other methods, such as so-called FORCs (*first-order reversal curves*), are technically and analytically more elegant—ultimately, a successful analysis of the rock magnetic signal will always be based on the combination of several methods and the combined interpretation of the respective results

5.1 Delineation from Paleomagnetism

Let's take as an example the mineral magnetite, often mentioned in Chap. 3. A ferrimagnetic[6] iron oxide—usually only micrometers in size, but an excellent recorder of the Earth's magnetic field—is present in (almost) all rocks of the Earth's crust in one variation or another. To track down this mineral, we take a sample of the rock under investigation and head to a laboratory with rock magnetic equipment.

What follows is a highly schematic illustration, applicable only in broad terms, of what is meant by *detective* work in the laboratory. Many of the rock magnetic measurements can be performed on the same samples that we used for the investigations of the Earth's magnetic field in Chap. 4. In fact, it is common rock magnetic practice to determine the underlying magnetic minerals of a rock's directional data to substantiate the quality of the results. In practical application, this also means that samples, so to speak in one piece, as they were taken from the outcrop, can be measured (but, as we will see, do not have to be).

In Chap. 3 it became clear that a mechanical destruction of the rock would also destroy any stored record of the Earth's magnetic field, as the remanent magnetic minerals of a sample would "tumble around" if a sample, for example, is ground. In rock magnetic investigations, which aim at a pure analysis of the type of magnetic minerals, samples can remain intact, but do not necessarily have to.

Attention, here comes a thought step to once again (Chap. 3) highlight an important point in the consideration of magnetic properties of rocks: In the storage of a magnetic direction, the storing magnetic mineral grain in the rock does not mechanically rotate in the direction of the field lines. The recording of the direction takes place, so to speak, internally—we remember (Sect. 3.1), that magnetism is very strongly anchored in the realm of electron movements at the atomic level.

When a direction is once stored and the mineral grain is mechanically rotated, the (internally) stored magnetic direction rotates with it. Since the recording of the Earth's magnetic field occurs not only in one mineral grain, but in many similar minerals in this rock, in the event of a mechanical destruction of the rock, all may possibly point in different directions when the grains tumble around, thus no longer as before in the direction of the recorded Earth's magnetic field. Since, as mentioned, laboratory-induced magnetic fields are used "anyway" in rock magnetic studies, the original rock composite is not necessarily required. It only needs to be ensured that the individual grains do not move relative to each other between inducing a field and the magnetic measurement.[7]

---

[6] Ferrimagnetics are a subgroup of ferromagnetics; there are several other distinctions and designations. A discussion is beyond the intention and usefulness for this book. See for example: Dunlop, D.J. & Özdemir, Ö., 1997. *Rock Magnetism, Fundamentals and Frontiers*, online publication 2010, Cambridge University Press, 573 pages.

[7] The easiest way to do this is to fill a container (a special magnetic sample box) completely with loose particles. In this way (using fields stronger than the Earth's magnetic field), it is also possible to use much smaller volumes of a rock sample (the size of a fingernail), which is good due to the design and underlying physical principles of some of the rock magnetic devices. For a DNA gene

For many geological questions, however, one will try to work with intact rock cylinders (i.e., as drilled from the rock association), if only to be able to use the samples for other non-magnetic investigations. In this respect, here are the first steps of our simplified laboratory investigations:

- The sample (intact rock cylinder) is measured with the same magnetometer that we used in Chap. 4 for the magnetostratigraphy studies. If a directional signal is measurable, i.e., the Earth's magnetic field was recorded, this leads to the first, trivial realization: Remanent magnetic minerals are present, i.e., such minerals, which usually occur in natural rocks with far less than 1% content. But we still don't know whether it is magnetite or another mineral or even a mixture of different remanent magnetic minerals. And certainly not the size or grain size distribution of these minerals. And these are precisely the pieces of information that allow conclusions to be drawn about the environmental conditions in the geological history of a rock.
- In a next step, heat the sample step by step to temperatures of several hundred degrees Celsius,[8] and observe from which temperature the remanent magnetization disappears. Magnetite has a so-called Curie temperature of around 580 °C (Chap. 3), beyond which the remanence properties disappear. At such high temperatures, even the ideal recorder magnetite cannot record the Earth's magnetic field. Other remanent magnetic minerals of the Earth's crust, for example hematite or goethite, have significantly different Curie temperatures. Some devices record the heating (and cooling) cyclically, in individual increments, and make it possible to analyze different minerals in a sample. Mixtures can also lead to unclear results, as explained before.

High temperatures can generally render the sample unusable for further magnetic, but also other geoscientific investigations, which is why pure heating is not the first choice for rock magnetic investigations in some cases.

Furthermore, the realization that it is a certain mineral is not sufficient. So we have to keep searching to characterize a mineral more closely.[9]

---

analysis, one would expect a tissue sample in the lab, not the entire person, let alone be able to measure it.

[8] Of course, there are special ovens/measuring apparatuses for this, which in particular can precisely realize temperatures above 600 °C and are shielded against any external, interfering magnetic fields.

[9] However, it should be noted here that some rock magnetic characterizations use not high, but on the contrary very low temperatures to identify a certain mineral based on an unusual change in its magnetic properties. Instead of an oven, our samples are then cooled down, for example, by liquid nitrogen or even liquid helium (for an idea, this is in the order of −270 °C) and measured magnetically.

5.1 Delineation from Paleomagnetism

In Chap. 3 we have already discussed that the grain size of magnetic minerals, like our magnetite discussed here, also has a significant effect on the magnetic properties of such a mineral. For many geological questions, it is often not enough to simply determine that it is magnetite; we also need to find out in which grain size or different grain sizes magnetite is present in our sample. Different grain size groups in the same rock often indicate from a geological point of view temporally separated processes of mineral formation and thus different—for example tectonic—environmental conditions in the geological development of a rock. In Chap. 3 we emphasized that in magnetostratigraphy we are looking for the recording of the Earth's magnetic field at the time of the formation of a rock. In dating, we are usually not looking for a generation of finest magnetic particles, which were formed at a later time after the original rock formation. These additional, later formed particles will after all be characterized by completely different geological processes than the minerals from the original formation phase of a rock. But if we do recognize later formed minerals as a separate group, we can also try to extract their geological information from them.

Another step in the detective analysis of our magnetic minerals could therefore be to gradually expose our rock sample to increasingly strong, laboratory-induced magnetic fields. The idea is as follows: Not only do different types of minerals—for example hematite versus magnetite—react more or less stubbornly to magnetic fields, but particles of different sizes of the same magnetic mineral are also influenced differently by magnetic fields.[10]

But more on that later. At this point, we want to remember the basic variation of the rock magnetic signal as a result of the geological backgrounds of a rock. This is easiest with a closer look at the magnetic susceptibility, and now is the time to read Sect. 3.5.

Before we turn to further subtleties of the geological interpretation of the sensitive magnetic ferromagnetics, we must open the black box of magnetic susceptibility as in Sect. 3.5, especially to better illustrate the change of the rock magnetic signal depending on the composition of the rock (Fig. 5.4).

---

[10] For larger particles compared to smaller ones, for example, stronger magnetic fields are needed for a reaction, which can then be read from a different curve shape with continuously increasing field strengths. One way to apply such fields, a so-called isothermal remanent magnetization (IRM), is comparable to a lightning strike—only better controlled in the laboratory in terms of direction and strength and without the visible and audible drama of the natural phenomenon of a thunderstorm. However, in rock magnetic practice, much more sensitive and often diagnostically successful methods are used, where not a step-by-step, but a continuous change in the strength of the applied fields is examined.

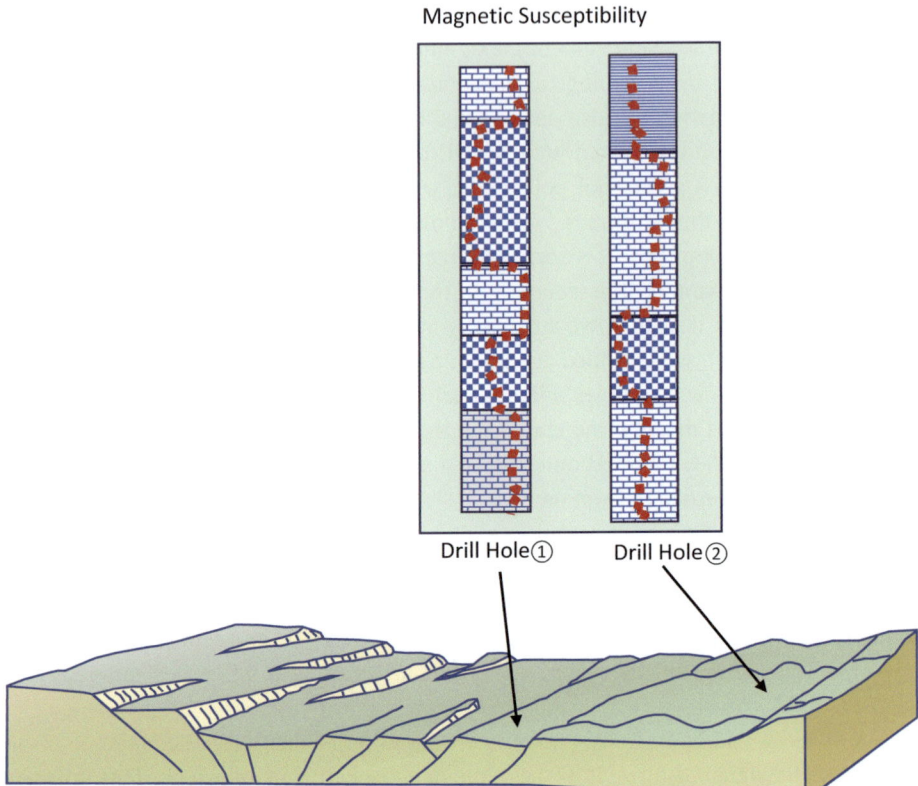

Fig. 5.4 Individual measurements of magnetic susceptibility (points) along the rock layers from a drill core or generally a sequence of layers (brick, check and line signatures) result in a pattern, which is controlled by the different composition of the rocks (see also Fig. 3.17, 3.18 in Chap. 3.5). In many geological scenarios, a comparison of the magnetic susceptibilities can serve as a proxy for the rock sequences in correlating different locations (here exemplarily drilling ① and ②). In Sect. 5.2 we want to understand this using a Buddha statue as an example

5.2 Applications of Rock Magnetism in the Geosciences

In the preceding chapters, especially in the geological interpretation and application of the magnetic barcode of the Earth's magnetic field, we have learned that exact (physical) laboratory measurements are extremely important in magnetic methods. We have also learned that for geological applications, when applied to geological sequences, particularly the recognition and use of individual patterns from these precise individual values are the key to success. In the following section, we elaborate on this claim with the recognition of typical (rock magnetic) patterns, thus being able to work more effectively with magnetic fingerprints than with the respective building blocks of such patterns. In

5.2 Applications of Rock Magnetism in the Geosciences

Chap. 4 we referred to a sequence of differently long "black and white" boxes (i.e. states of the Earth's magnetic field) as a pattern, while individual black or white boxes have only limited significance on their own. In the following, we create with rock magnetic measurements other building blocks from which we assemble our patterns for the interpretation of our Earth.

Susceptibility and a Buddha

A rather unusual application is found for the fingerprint of the rock magnetic minerals in the destroyed giant Buddha statues in the Afghan Bamiyan Valley. This chapter does not attempt to summarize or even describe in their complexity the extensive, years-long work of various international organizations, national institutions, or individual participants, let alone comment on them.[11] Rather, the example of a thick sequence of sedimentary rocks in the Hindu Kush is intended to illustrate the relationship between the changing deposition conditions of these rocks and their representation in the magnetic properties of these rocks. The specific application example of the Buddhas hopefully makes the following explanations in this respect more accessible.

Our presentation is a typical example of a magnetic pattern matching, which, as introduced earlier, takes advantage of the principle of using the rock magnetic signal as a proxy for "geological properties". Accordingly, a methodology for using magnetic susceptibility is presented below, with which the role of geology in the reconstruction of the Buddhas could be answered and tested.

After the destruction of the Buddha statues in the Bamiyan Valley in Afghanistan in 2001, it was necessary to solve a puzzle that presumably resembled a broken sculpture. At least two enormous statues had been carved into a natural sedimentary rock cliff on the border of the Hindu Kush with the Koh-i-Baba mountains, 230 km northwest of the capital Kabul, in the 6th and early 7th centuries, about 800 m apart. The larger of the two statues was approximately 55 m high, 18 m wide, and just as deep. An internal side entrance led from the lower chambers to a viewpoint above the Buddha's head, overlooking the Bamiyan Valley, which is sitting at about 2500 m altitude[12] with its 5000-meter peaks of the surrounding mountains in the background (Fig. 5.5).

---

[11] From the multitude of relevant publications: *The Giant Buddhas of Bamiyan, Safeguarding the remains*, 2009, Petzet, M. (ed.), International Council on Monuments and Sites ICOMOS, 280 pages, Bäßler Verlag, Berlin; *The Future of the Bamiyan Buddha Statues. Heritage Reconstruction in Theory and Praxis*, 2020, Nagaoka, M. (ed.), UNESCO Publishing. Springer. Open access.

[12] At its peak, the center of the Gandhara school of Buddhism was located in the Bamiyan Valley and along the bustling trade route of the Silk Road. One can imagine a scene during this golden age where the possibly partially gold-painted and fire-lit Buddhas were widely visible, surrounded by several hundred internally richly decorated monk caves in the same cliff. A place of gathering, spirit, and contemplation. A place that has remained of spiritual significance to many millions of people to this day.

Fig. 5.5 A partial view of the nearly vertical cliff in the Bamiyan Valley, oriented approximately west-east, shows the niche of the larger, western Buddha. This photo from 2006 shows the blasted fragments of the Buddha at the base of the niche. The nearly horizontal layering of the sedimentary layers in this area of the Hindu Kush (Fig. 5.6) is particularly traced at the base of the cliff by numerous caves once built by monks. Erosion by wind and occasional rainwater has morphologically changed the cliff over geological time. The deep, vertical cuts into the cliff testify to the advanced and constantly progressing erosion of the rock. A mental comparison of the gentle morphology of the mountain peaks in the background with the rugged, sharp-edged (metamorphic) rocks of, for example, the European Alps[13] impressively emphasizes the softness of the cliff's sediments[14]

With the international outcry over the destruction of the Buddha statues, UNESCO, in collaboration with ICOMOS (International Council on Monuments and Sites), subsequently took action to safeguard the Bamiyan World Heritage Site. It quickly became

---

[13] For comparison: The Matterhorn on the border of Switzerland and Italy is just under 4500 m high.

[14] In the rock association and even as pebble-sized fragments, the sediments have rock hardness. However, when placed in water, a pebble-sized fragment dissolves into its components, i.e., individual grains, within a few minutes. As a rule, the individual grains are merely interlocked by overburden, which results in the hardness of the rock. Often sedimentary rocks have a so-called binder, formed from a carbonate or silicate solution of circulating waters, which cements the grains together. Such a binder is missing here and explains the predisposition of the cliff's sediments to erosion, but also the good suitability for sculpting the rock.

5.2 Applications of Rock Magnetism in the Geosciences

clear that in addition to restorers, architects, and engineers, geologists in particular would be needed to understand the original position of the numerous fragments. After all, most of the fragments, in a very pragmatic, secular view, are rock fragments from the face of a cliff made of several million-year-old sediments.

Would it be possible to reassign the fragments of the blown-up statues to their original position, perhaps even rebuild the entire statue or at least follow the architectural concept of anastylosis for a partial reconstruction? How would such an undertaking be technically feasible? Rock magnetic investigations cannot and should not answer all questions in dealing with the destroyed Buddhas—but a significant detail, the question of the correct original position of the recovered fragments, could very well be contributed to. Such is, initially at least, the theory, and this approach is to be explained here. As said, especially to illustrate the principle of working with magnetic patterns in environmental magnetism.

As we have already stated, each rock, depending on its mineral components, contains a typical rock magnetic signature. Very similar to a genetic code or a fingerprint—completely independent of the existence or recording of the Earth's magnetic field—each mineral that makes up a rock at a certain location contributes to its magnetic susceptibility (Sect. 3.5). No age can be read from such a signature alone (Sect. 4.1), nor is there a direct indication of the origin or the underlying composition of the rock (Fig. 3.17). However, by comparing the signatures of two rock samples, their regional or local neighborhood can very likely be determined based on relatively similar characteristics of magnetic susceptibility (Fig. 5.4). If we consider the susceptibility values as a proxy for the change in the composition and texture of the rock itself.[16] Even fragments from a rock wall still contain geological information. If these rock fragments are small, the recording of geological variations must therefore be correspondingly high-resolution.

To understand how this can specifically help us in cleaning up and assigning the shattered Buddha fragments, or indeed in any other form with the correlation of geological information in purely natural deposits of earth history, here is an analogy from everyday life: When a valuable vase breaks, we try in a first step towards reconstruction to collect as many shards as possible (Fig. 5.7). However, this will not always be completely possible for events that occurred a long time ago. The individual pieces of the broken treasure can then be put back together in a second step, similar to a conventional picture puzzle, and with a little effort. We will generally use the following criteria for assigning the shards:

---

[16] In Fig. 5.4 no scale is given to suggest the thicknesses of the geological profiles proposed for comparison. The reason for this apparent omission is, upon closer inspection, a huge advantage of the method: If the susceptibility measurement points are very close to each other, the compared profiles can also be short, even in the centimeter range. To compare profiles of hundreds of meters in thickness, it is sufficient to measure the susceptibility at meter intervals. In both cases, it is important not to skip mutually significant layers during the survey, which would create an artificially altered fingerprint.

Fig. 5.6 Compared to Fig. 5.5, a pivoted view of the western Buddha niche. Drawn white auxiliary lines and the nominal division of the cliff into sediment packages A to E illustrate the lateral extent of geologically uniform sediment layers. Package D represents a thickness of about 40 m and thus serves as a scale in this picture. The packages A to E were arbitrarily chosen, mainly to illustrate the two-dimensional horizontal layering. A geological zoom on the cliff would reveal a multitude of further, only a few centimeters thick sediment layers as finer subdivisions of the packages A to E (Fig. 5.15 and 5.17). This latter perception of the layering is important for understanding the applied magnetic methodology. The sediments, with an age of a few tens of millions of years, are geologically young and, although solidified by overburden and compaction, are still easily eroded by wind and occasional rainwater. As witnesses of this erosion, some debris cones are drawn in the top and bottom area of the cliff. These taluses[15] partially cover the uniform layering

1. The shape (outline) of the shards (comparable to the outlines of a puzzle piece)
2. The form, in the sense of a curvature of the shards due to the original position in the vase (here a vase differs from a conventional picture puzzle due to its three-dimensional shape)
3. The decoration, if present (very much in the sense of a picture puzzle)

---

[15] "Talus" (singular) is a geological term for (rock) debris cones.

5.2 Applications of Rock Magnetism in the Geosciences

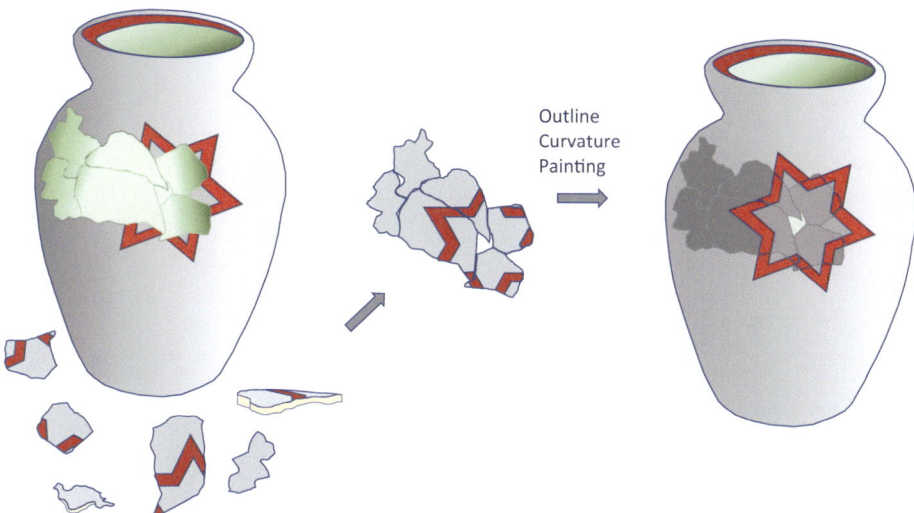

Fig. 5.7 In everyday life, we use features of the fragments, such as shape, curvature, or even painting, to assemble broken objects, for example, a vase. Since such criteria cannot be used in the case of the Buddha statues, we try to create alternative characteristics for each fragment through magnetic susceptibility. The scenario presented is just one example for any kind of geological investigation in which rocks of the same sequences are to be correlated with each other

The success of the reconstruction depends on the completeness of the available shards and, in the—at least aesthetic—sense, the suitability and skill in handling the adhesive used. Even with complete shards, gaps remain primarily where the vase has disintegrated into such small and smallest fragments (in the extreme case to dust) that an assignment of these components to the original position is no longer possible. An example is the glaze at the edges of the larger pieces. In this case, none of the required features—outline, shape, or painting—can be used for reconstruction.

It is easy to understand that additional information, such as photos or drawings of the vase, can be extremely helpful in the reconstruction if, on the other hand, all other features required for reconstruction—shape and painting in this case—are available without further effort. We certainly do not need rock magnetic methods or other scientific methods as aids for a reconstruction in such cases.

The case with the Buddha fragments is different. Features such as external painting or sculpted elements are not or rarely present in the fragments. The shape—both in terms of an outline and a curvature—will hardly be helpful with the rock fragments. Especially with soft sediments (as in the present example of the Hindu Kush), hardly any identifiable fracture surface will be available. Especially when the fragments are transported away from their original position, whether in the present case by the explosive force of the blast, or by natural erosion at a cliff, the fracture surfaces will subsequently be altered by abrasion. In most cases, only the—all around—worn core of each fragment

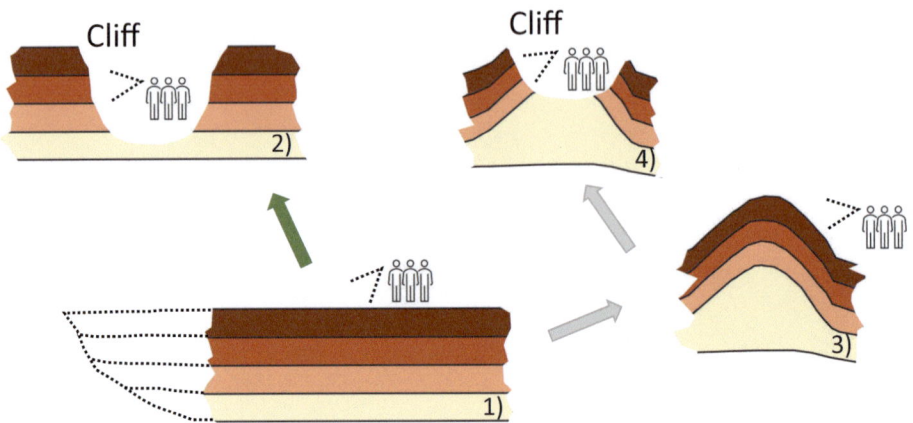

Fig. 5.8 Explanation for the use of the term "cliff" from sedimentary rocks. Successively, horizontally deposited layers (**1**) are eroded, for example by a river, and thus cut laterally (**2**). Possibly the layers are first deformed (**3**) and then cut (**4**). In situation (**1**) and (**3**) only the youngest layer can be considered; only in situation (**2**) and (**4**) the older layers become visible through the cut. The initial deposition of the layers could, for example, occur similarly to Fig. 4.4

remains as a puzzle piece for the reconstruction. The two sides of a separating surface have thus become unrecognizable to each other through individual changes.

To better understand the close relationship between the Buddha statues and geology, let's first take a look at the cliff in the Bamiyan Valley. Our main focus is again on understanding the use of the rock magnetic signal as a proxy for a geological sequence. Most of us will have an idea of the term "cliff". Nevertheless, we want to emphasize a significant point with Fig. 5.8: From a geological point of view, the often vertical walls of a cliff reveal a lateral section of the otherwise hidden geological layers.

Sediment layers are deposited sequentially, horizontally, and with a finite lateral extent, following a geological basic principle. This is shown by the lighter to darker coloration in the schematic representation of Fig. 5.8 (1). For a better understanding of the term "finite lateral extent", imagine filling a bowl in the kitchen layer by layer (indicated on one side by the dashed lines). In our geological case, such a "bowl" would have a finite size.

Erosion following the deposition and solidification of the layers, for example by a cutting river or a glacier, occurs from the upper youngest layer to the lower older layer (2). In this way, a cliff is successively created, which reveals the layers.

In an alternative geological scenario, the sediment layers are initially deformed (3), for example in a mountain formation process in the plate tectonic sense. An observer only looks at the top of the youngest layer.

Under (4), it is shown how a cliff has also been created via the intermediate stage of layer deformation. The cliff is created, as explained under (2), by erosion. Many other geological scenarios for cliff formation are conceivable. Most such geological processes

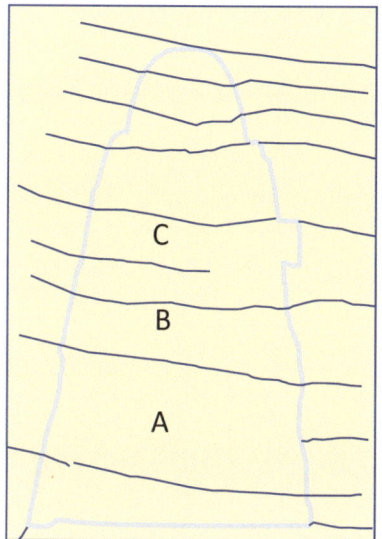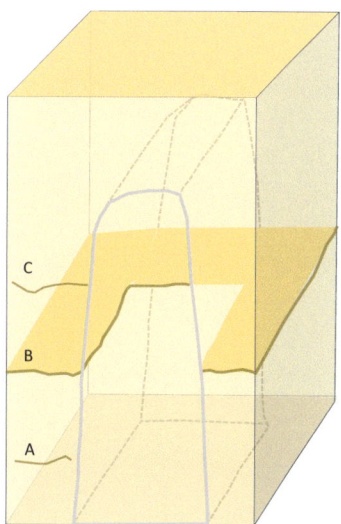

Fig. 5.9 Front view of the cliff, at a time before the niche and Buddha were sculpted from the rock. The horizontal lines indicate, as in Fig. 5.5 and 5.6, the almost horizontal layering of the rock layers. The three layers A, B, and C are highlighted for orientation. The outline of the future Buddha niche—not the Buddha figure itself—is indicated by a line. The cuboid on the right in the picture illustrates the three-dimensional layering of the sequences of the cliff, as is also assumed in Fig. 5.10 ff. The dashed lines in the right picture suggest the future niche of the Buddha chiseled into the cliff from a perspective view

are significantly more complex and go through several phases of deformation and erosion. However, the basic principle should suffice for explanation in order to assess the geological situation in our exemplary Bamiyan Valley.

People would have found such a vertical wall more than 1.5 millennia ago when they began sculpting the Buddha statues. The main steps from the initial phase to the blasting are explained in Fig. 5.9, 5.10 and 5.11, and summarized in the following three points. Essentially, our considerations should show that the blasted rock fragments originate from a three-dimensional, geologically understood rock association. A fragment thus becomes comparable in a pragmatic view to a core sample from a deep drilling or a geological handpiece from an outcrop.

- A vault-like niche is chiseled out of the soft sedimentary rock of the cliff, with the sculpted semi-form of the Buddha being "left standing" in the "rear" area of the niche.
- The blasting destroys the semi-form of the Buddha and leaves a now almost flat, vertical rear wall in the rear area of the niche.
- The fragments of the Buddha mostly lie in the lower foot area of the niche, while the niche itself remains essentially preserved as a hollowing out of the geological cliff.

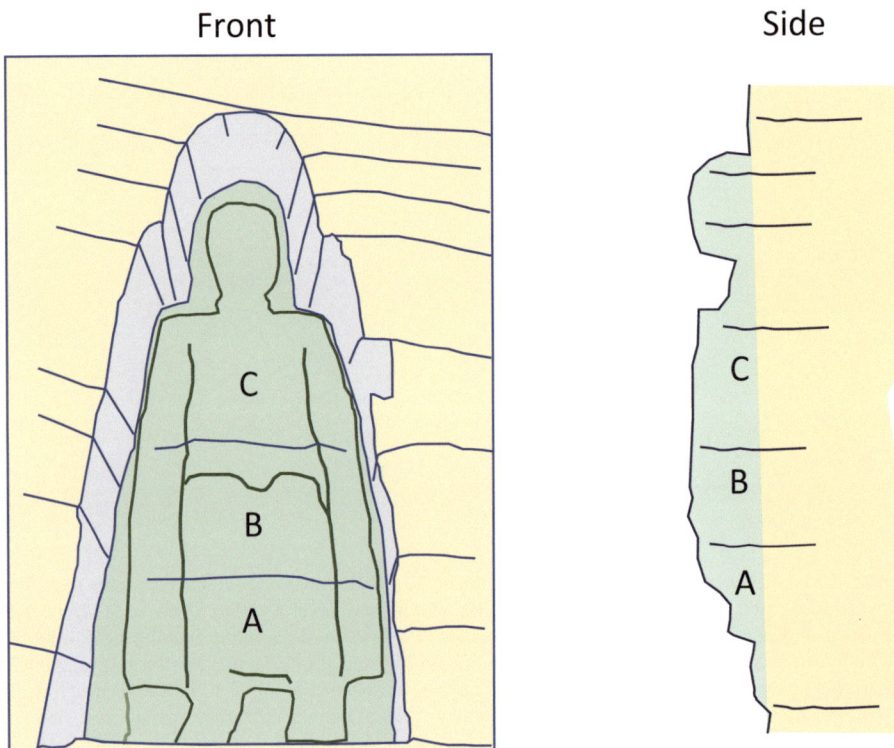

Fig. 5.10 Schematic representation of the Buddha carved as a semi-form from the cliff in its niche. The right illustration shows a theoretical side view to illustrate the relationship of the sculpted areas with the rocks of the cliff. Except for possibly subsequently applied, thin clay layers for "dressing" the sculpted layers, the statue consists of the in-situ rock of the cliff. The left illustration clarifies that the side walls allow a view of the three-dimensional layering of the schematized layers. Highlighted are three theoretical rock units (A to C). Since we want to use the rock magnetic signal of the layers as a proxy for the geology in the following, an understanding of the three-dimensional structure is necessary for understanding the procedure

The natural, horizontal layering of the rock strata remains unchanged in the now sculpted Buddha (slightly darker shading in Fig. 5.10 ff.). Particularly in the "interior" of the several meters deep semi-form, the natural sequence of rocks has not been altered in any way. Only the surface may have been covered with a thin—centimeter-thick—layer of clayey material[17].

---

[17] From a rock magnetic perspective, subsequently applied material should be avoided in the measurements, as we will easily understand in the further explanation of the methodology used here. Only natural rock is to be used for the question presented here, but then it is no problem whether this rock was possibly altered in the form of one of the blasted fragments and possibly additionally by sculpted elements on its surface. Unless this form was covered with artificially applied clay.

Fig. 5.11 The blasting of the Buddha leaves a nearly flat, vertical back wall of the niche. The sculpted area has broken away from the back wall. The majority of the Buddha's fragments—from tons-heavy boulders to dust particles—now form a debris pile at the base of the niche. Some fragments were hurled out of the niche into the valley. The fragments were gradually recovered in the following months during the work of UNESCO and ICOMOS and were then also accessible to magnetic measurements

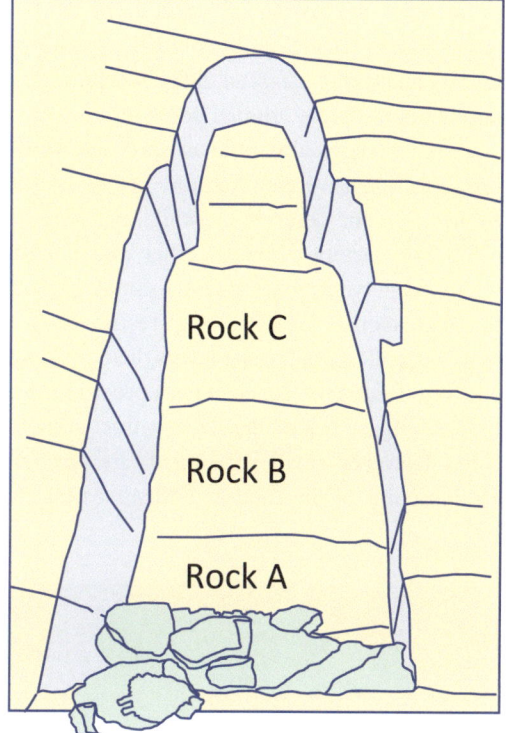

From a constructional point of view, while somewhat ignorant artistically speaking, the hollowing out of the Buddha niche—with its vertical side walls and the vault-like roof area—and the sculpting of the Buddha—which thus forms the back wall of the niche (Fig. 5.10 right)—could be seen as one and the same process.

Unlike the vase mentioned earlier as an example—a hollow form—a considerable part of the fragments will come from the interior of the Buddha and thus cannot contribute any external form features per se. Geologically, these are still soft, almost hand-wipeable sedimentary rocks[18], and the preservation of significant fracture surface against the respective neighboring fragments is unlikely. The glaze, to return to our vase analogy, is almost everywhere broken off. Shape, curvature and painting are not usable.

Nevertheless—and this is the essential point of our thought experiment at this point—the analogy Buddha/vase/puzzle remains valid if it is possible to replace the criteria of shape and painting in the shards of the vase with other features tailored to the specific situation. Since the Buddhas are made of natural sedimentary rocks, it is obvious to look for suitable criteria in the geological properties of the rock layers. If it is possible to

---

[18] In the main, these are clays, silts and sands. These are geological terms for very fine to easily visible grain sizes of the sediments.

characterize the geological structure of the respective fragments of the Buddhas and to put them into a meaningful context—in other words to reconstruct the depositional space in the area of the Bamiyan cliffs in space and time—the fragments can also be put back into their respective original position.

Purely geologically, such an approach would be possible, but extremely time-consuming, labor-intensive and therefore costly. Due to the similarity of the deposits in the area of the cliffs, the description would also have to be resolved into layers of a maximum of a few centimeters thickness. This would hardly be possible without additional at least microscopic work. For this, rock samples would have to be transported in large quantities to (international) laboratories to realize such investigations. It seems obvious to use the rock magnetic signal of the layers, here the magnetic susceptibility, as discussed before, as a more effective replacement (as a proxy) for a classical geological approach (Sect. 3.5).

The largest fragments have a three-dimensional size of about 2 m (Fig. 5.12 and 5.13). However, smaller fragments are much more common and range down to the individual grains of the sand and siltstones that make up the cliff. After the explosion, the

Fig. 5.12 The picture gives an idea of the size range of the fragments (note the scale). In the background on the right, part of the sculpted surface of the Buddha can still be seen on one of the larger fragments. If the sculpted surface is covered with an artificial clay layer, this artificial layer cannot be used for the magnetic measurements, but forms a valuable orientation surface in the sense of the puzzle

5.2 Applications of Rock Magnetism in the Geosciences

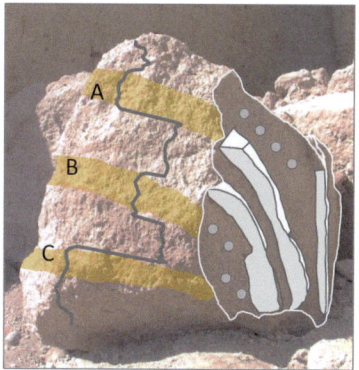

Fig. 5.13 The left picture shows one of the larger fragments with an edge length of about 2 m. In the right picture, three (arbitrary) layers A, B and C are visually highlighted on the front fragment surface; the sculpted side surface with possible remnants of an artificial clay layer is also graphically emphasized. In order to measure a meaningful profile of the magnetic susceptibility, which runs perpendicular to the layering, an overview description of the geological features of a fragment is made for each fragment. The dark, fictitious line perpendicular to the layering symbolizes a possible result of the measurements of the magnetic susceptibility, as it could result from measuring points at centimeter intervals. At the corresponding place in the still remaining rock of the back wall of the niche, the same geological layers and thus the same magnetic measurement signal are expected. In the case of the fragment shown, the sculpted surface helps with spatial orientation and can be used as additional confirmation when positioning the fragment in relation to the Buddha. Magnetic measurements of the clay layer on the surface, on the other hand, are avoided in the magnetic measurements, as of course no correspondence in the geology of the back wall is to be expected. So here is the principle of the magnetic method for repositioning the Buddha fragments. We are looking for the same pattern—the same variation of the magnetic susceptibility over a measuring distance perpendicular to the layering of the sedimentary rock—in the back wall of the niche and in a respective fragment. Similar to the methodology in Sect. 4.1 and the comparison of the magnetic barcode with the standard time scale, a significant pattern can only be formed from a certain minimum number of relative variations in the signal of susceptibility

rubble piles up mainly in heaps at the base of the niche.[19] No one will consider returning individual grains to their original position. However, the question of the minimum size of a fragment to be reconstructed or used in the sense of an anastylosis is of course quite justified. In such considerations, criteria for the clear identification of a fragment (in the case of a vase, for example, shape and painting) are theoretically in the foreground, but must certainly also be balanced with criteria for the practical feasibility of a reconstruction. This consideration is most likely also valid for the more general application in purely geological questions—if only in terms of the appropriate order of magnitude, in the sense of the resolution of a geoscientific study. Who would expect exact uniformity

---

[19] For those who wish to follow the analogy with a purely geological study, for example, in the correlation of two drill cores (Fig. 5.4) in mind: We can compare the blasted fragments with the more or less broken cores of a geological drilling. Measurements and investigations of these core pieces, to correlate them with the surrounding geology, can then indeed be compared with the repositioning of the fragments.

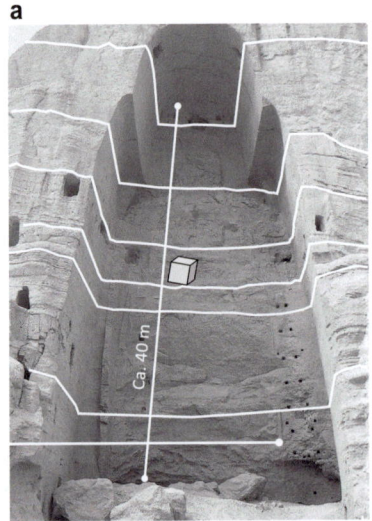

Fig. 5.14 a) The 2006 photograph of the niche of the western Buddha visualizes, in addition to the horizontal layering of the geological strata (bright lines in the left picture and surfaces in the right picture), especially the dimension of a fragment of about 2 m edge length (cube in the center of the picture on the left) compared to the about 40 m high niche. Note, the foot area of the overall higher niche is still filled with rock debris. Another comparison for understanding the volume of blasted rock is made in the text. b) Drawing of the empty niche of the western Buddha in Bamiyan. The sketch on the right shows, using the example of a single sediment layer (continuous line) and a series of cuboid-stylized and lined up fragments with an assumed edge length of about 1–2 m, how the rock blocks were theoretically positioned in a layer in the cliff's bond. Depending on the horizontal position in relation to the semi-form of the Buddha, a larger or smaller number of fragments is to be assumed both laterally and in depth, corresponding to the slimmer or wider areas of the figure. Due to the destructive force of the explosion, it would also not be assumed that an equal number of fragments with an edge length of 1–2 m has remained for all horizontal sediment layers

of a sandstone, for example in the millimeter range, i.e., the exact same grains in the deposits of a river in the millimeter range? This may seem quite unlikely without further consideration. Rightly so.

> When we consider the enormous volume of rock of the Great Buddha, sculpted as a semi-form from the sedimentary cliff[20] (Fig. 5.14a, b), it becomes understandable that, in addition to the sheer amount of fragments of all sizes, there is an extremely small proportion of fragments with sculpted surfaces, i.e., surfaces that can be used as features in the sense of the broken vase analogy. In terms of a possible reconstruction most of the fragments are, quite soberly and pragmatically considered,

---

[20] Another striking comparison: The niche of about 55 m x 18 m x 18 m corresponds to 17,000 m³. At 2 t weight per cubic meter, this amounts to 34,000 t of rock material. If about half of this was present in the Buddha statue, that's approximately 5% of the Empire State Building in New York.

5.2 Applications of Rock Magnetism in the Geosciences

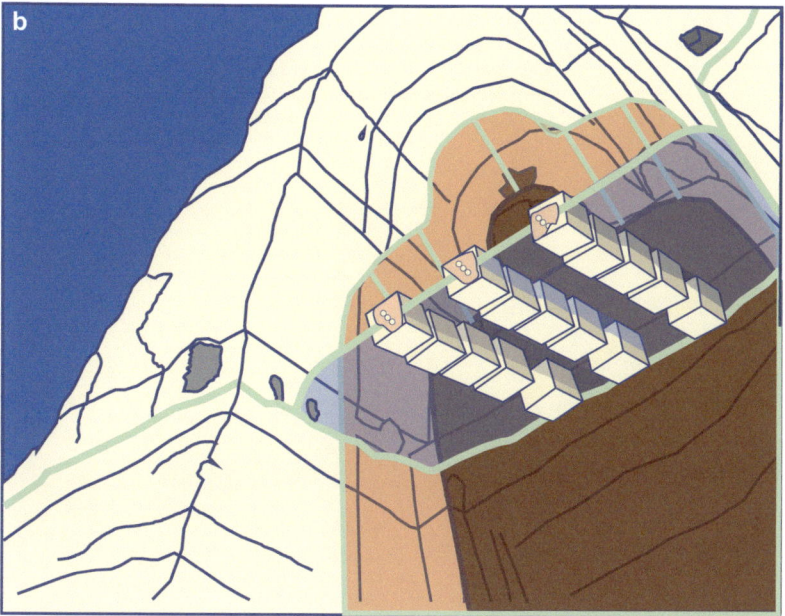

Fig. 5.14 (continued)

> and while indeed fragments of a unique work of art, characterized by nothing else than they are pieces of rock or even grains of sand: geology, more precisely, the natural features of a rock—features which reflect the formation process of the respective rock in their type, size, shape, and arrangement of components.

At several points in this book, we have considered in this sense that it is always worthwhile for scientific studies to align the level of detail with the question at hand. Such questions include, for example, in Chap. 4 the necessary sample density to not miss any of the recorded polarity changes of the Earth's magnetic field in different geological scenarios with different accumulation rates of the sediments. The solution for the optimal sample density was to create a compromise between the shortest duration—expected according to the GPTS—of the polarities and the speed of the accumulation of the sediments, i.e., how long it took to accumulate, for example, 1 m of new rock.

If we now, as suggested here, approximate the geological stratification by measuring the magnetic susceptibility, we must ask ourselves in a very similar way, how many (parallel) magnetic profiles we should probably measure (Fig. 5.15 and 5.16) to adequately capture the geology. If our assertion is correct (Fig. 5.5 and 5.6) that the sediments are horizontal and still almost tectonically undisturbed rock layers, obviously a single profile would likely be completely sufficient. To verify this, we approach the back wall of the Buddha niche a little closer with Fig. 5.15 and look at the succession of the layers in more detail.

Fig. 5.15 Partial view of the back wall of the niche. The white lines trace—as already in Fig. 5.6—the geological layers. Viewed from a distance (or averaged over unevenness), the layers lie horizontally and parallel. As shown here, in detail, typical characteristics of sedimentation at the time of the rock's formation are revealed: layers—in the centimeter to meter range—vary in lateral thickness, sometimes wedge out completely, i.e., are present in one area of the back wall, but not in a corresponding area to the side. Examples to explain variations in thickness or the lateral end of a layer include the dynamic interplay of deposition and erosion in a meandering river or sedimentation on an "irregular" relief

The possible reasons for the lateral variation of a rock layer as in Fig. 5.15 are extremely diverse. Among other things, the mineral composition or the grain size distribution of a sediment can change, but a complete discussion is beyond the scope of this book. As an example of lateral variation within a sediment layer, consider the simultaneous deposition in different areas of a river bed. Due to the different flow velocities, the coarsest particles of the sediment load are still transported at the central flow area of a river bed, while, for example, in the quieter flow shadow in the river bank area, even fine grains come to rest and are deposited.[21]

---

[21] The meanders of a river winding through the landscape are a fitting image for different flow velocities also at the river bed: A paper boat on its way downstream will always follow an ideal "path" from outer curve to outer curve, because there the speed of the flowing water is highest. In the inner curves, the boat might possibly get stuck or finer sediment material comes to rest.

5.2 Applications of Rock Magnetism in the Geosciences

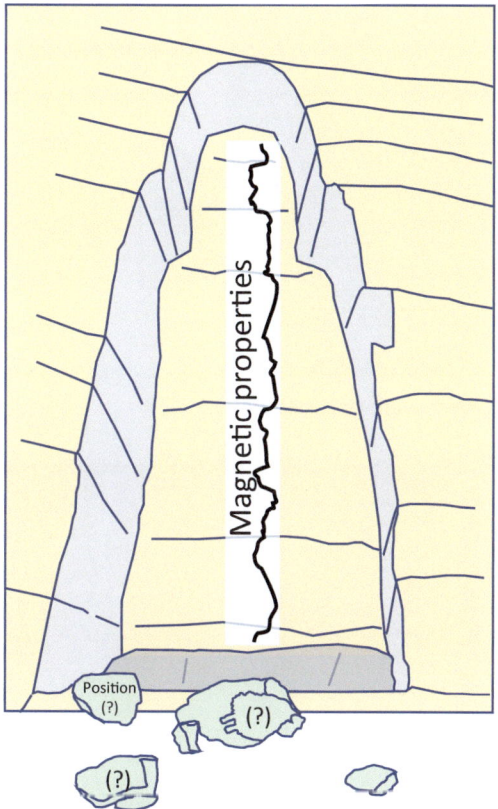

Fig. 5.16 When all fragments have been recovered, the question as to their original position—as components of the Buddha in the cliff—remains to be answered. The undisturbed, horizontal layers of the cliff allow their characteristic properties to be recorded with a few (ideally one) geological profiles running perpendicular to the stratification. The text explains the advantages of describing the sediment layers with a series of magnetic measurements (here the magnetic susceptibility), instead of subjecting them to a more elaborate, classical geological rock description. Sufficiently large fragments are subjected to the same type of magnetic measurements and are then to be assigned to the original position in the rock wall (ergo of the Buddha or more generally formulated of a comparable geological sequence of layers) via a pattern comparison

It remains to be noted: Some reasons for the lateral variation of rock layers can lie in the depositional environment. While various media such as wind or water transport and deposit rock material in different ways, this usually happens in discrete zones. In this way, lateral differences within a sediment layer are created, or this layer *tapers* laterally (another interesting geological expression; Fig. 5.17), with the lateral continuation being made by a—younger, later deposited—layer. Anyone who is interested, but does

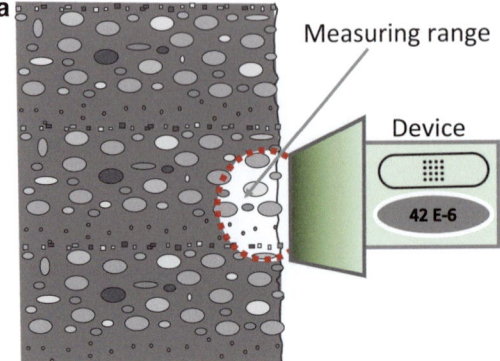

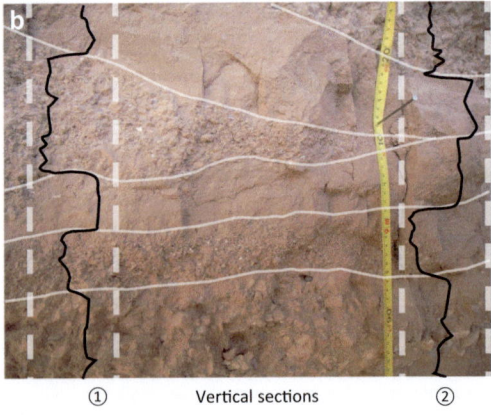

Fig. 5.17 Further partial view of the back wall of the niche with a vertical height of 40 cm (**b**). In the lower area, the few centimeters thick layers (white lines for emphasis) lie with only slight variation approximately parallel; especially in the upper 15 cm of the image, the wedging out of a complete layer is clearly recognizable. Two vertical profiles (① and ②) are superimposed as fictitious measurement values of the magnetic susceptibility on the photo (black lines). To perform such susceptibility measurements, a value is determined with a portable magnetometer (**a**) along the vertical line of profiles ① and ② at centimeter intervals. These measured values are plotted against the profile height to obtain the curve (black line) of susceptibility. Note the similarities (lower half of the profiles in **b**) and variation in the measured values depending on the layer boundaries (cf. Fig. 5.18). The absolute value of the susceptibility of a single measurement should be correctly indicated in a suitable physical unit. Contrary to scientific convention, we omit a value axis in the figure to emphasize the importance of the similarities and differences in the pattern of the measured curves of profiles ① and ② compared to the individual values that make up the curves. We have explained in more detail in Sect. 3.5 what a susceptibility value in rocks results from. We can emphasize at this point that it will not be possible to distinguish the sediment layers of the cliff from each other by a discrete value. It would be handy to determine a value of, for example, 42 susceptibility units and to be able to conclude unequivocally from this to a light sandstone with a low matrix content in the middle area of the cliff. Unfortunately, nature is too diverse for this, or in other words, the possibilities of determining the same susceptibility value for the most diverse rock compositions are too multifaceted. The variation of successive measured values over a measurement profile (see profiles ① and ②), the resulting pattern of the same and in the same way successive rock layers, will however be a typical and significant pattern for the examined depositional environment. After all,

◀ we use the magnetic susceptibility as a proxy for the rock. The rule is that the higher the resolution—here we mean the distance between measuring points be it in the millimeter or ten meter range—of such a pattern resulting from corresponding measured values of the susceptibility is considered, the smaller the area will be, geologically speaking, in which such same deposits can be expected. Furthermore, to (a) A suitable measuring device determines the magnetic susceptibility as the sum of all components in a certain volume of rock. The components—in sediments these are for example quartz grains, clay particles, shell fragments with varying proportions—are captured over a circular area (let's say, 2 cm in diameter) and, like a hemisphere, up to a few centimeters into the depth of the rock wall. Results are given as the sum of these individual values of the magnetic susceptibility at measuring position X.

not want to delve deeper into the diverse world of sedimentology,[22] might want to pay attention to the movement of the washed-over sand grains with incoming and outgoing waters on their next beach visit—or make a paper boat.

To emphasize the importance of searching for such magnetic patterns, absolute value axes are either omitted in this chapter or represented by relative statements such as "weaker or stronger" (Fig. 5.19). Once we have understood and accepted the principle of pattern matching, we can open ourselves to the limitation that absolute values of susceptibility in the adequate physical unit are **not at all** irrelevant or without any diagnostic value: Some rocks reach values of susceptibility strength that could never be reached by other rocks due to their typical mineral components. The order of magnitude of the measured values thus takes on a diagnostic character (Fig. 3.17). A sandstone consisting only of quartz will have susceptibility values lower by orders of magnitude than a basalt enriched with the finest magnetite components. If two such different rocks are in direct proximity in a geological profile (which is not impossible), the clear discrepancy in values results in a likewise clear, easily recognizable, and thus assignable magnetic signature. In Fig. 5.18, several such sharp transitions between different types of rock are shown. As a pattern, we want to understand the totality of variations in comparison from one profile to another. In the majority of geological rock sequences, we will embark on the search for such patterns in a sequence of sedimentologically similar rocks without sharp layer boundaries. The search for relative changes under the label "weaker or stronger" is then all the more relevant.

The schematic representation of the profiles in Fig. 5.18 illustrates the variation of magnetic susceptibility with the sediment layers, as is to be expected when measuring the back wall, but also the fragments. The absence of individual layers in comparison between profiles ① and ② in Fig. 5.18 has the most significant influence on differences in the respective pattern of susceptibility. The light lines between the profiles connect correlating points in both profiles. Differences in thickness within a layer lead to inclined correlation lines (Fig. 4.17). The scheme can be applied to profiles in the centimeter range as shown in the example, but of course also retains its validity on a larger scale,

---

[22] For example, Leeder, M.R., 1982. *Sedimentology, Process and Product*. Chapman and Hall. Springer Science & Business Media, 2012.

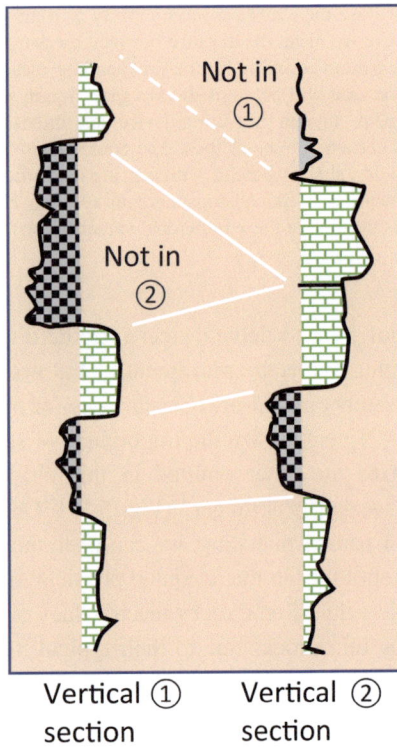

Fig. 5.18 Schematic correlation of two fictional sediment profiles based on relative variations in magnetic susceptibility due to a changing composition of the rocks (checkered and brick signatures) from layer to layer (see Fig. 5.15 for an explanation of why not all layers have to be equally thick or did even develop similarily in both profiles). In this example, the correlation is primarily due to the clear changes at the lithological boundaries. The same principle of correlation based on the pattern of susceptibility applies within a homogenous lithological unit, for example, a thick sandstone package

for example for layers with thicknesses in the meter range. For each measured value of magnetic susceptibility, an absolute (numerical) value can be given in a suitable physical unit.

Our example, as explained, does not show a corresponding value axis: The relative variation of the susceptibility measurements is important. The pattern from lower to higher successive values is considered and highlighted, not the individual measured values.

For our methodology, we will want to find the same pattern in the measurements of an assignable fragment (Fig. 5.19) and the measurements of the rock in place on the back wall (Fig. 5.20). This assigns a fragment to its original position, just as a puzzle piece can be inserted into the only possible place in a puzzle. The small-scale variation of the stratification (Fig. 5.17) implies the following for the application of magnetic susceptibility as a proxy for sediment layers: A profile perpendicular to the stratification in the middle of the back wall of the niche (Fig. 5.16) will not yield exactly the same result as a profile a few meters to the right or left of the center, at least not for all measurement points, if these are taken at centimeter spacing.

However, the overarching measurement course of magnetic susceptibility will be quite similar, much like our example of sediment layers, which appear flat and parallel, but show more detail upon closer inspection. Consequently, we must decide whether a

5.2 Applications of Rock Magnetism in the Geosciences

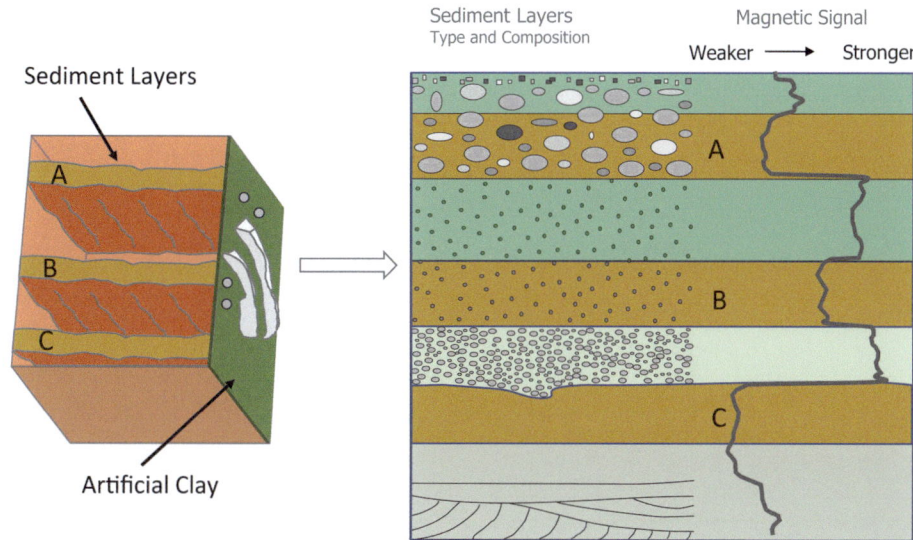

Fig. 5.19 On the left, the fragment from Fig. 5.13 has been stylized into a cuboid with the three exemplary layers A, B, and C to illustrate the three-dimensional orientation of the layers compared to the possible artificial clay surface in the fragment. The space between layers A, B, and C was naturally also filled with sediment in reality, but is here omitted to better illustrate the orientation of the layers. On the right, the variation of magnetic susceptibility (magnetic signal) as a function of the varying composition of the respective sediment layers has been schematized. A different composition of the sediment implies a different strength of magnetic susceptibility—for the methodology presented here, it is not the absolute value at a certain point of the curve that is decisive, but the pattern that results from the respective change from lower to higher values and vice versa along a measuring track (see also Fig. 5.18)

profile across the back wall is sufficient or whether several adjacent profiles are necessary to meaningfully represent the geology. This decision can only be made through a geological overview description[23] of the layers in the area of the back wall of the niche. Similarly, in the general case, it is usually not sensible to undertake purely geological investigations without understanding at least the regional geology.

In this discussion of the application of the rock magnetic signal in geology, we have not only approximated the geological range of rocks by measuring the magnetic susceptibility, but especially learned about the special importance of working with patterns or rock magnetic fingerprints outside of the GPTS. As an analogy for the re-positioning of the Buddha fragments, we replaced missing, more traditional features such as shape, painting, and curvature of shards of broken artworks by a pattern of varying magnetic properties of rock fragments. It's as if we were using fingerprints on the glaze instead of the usual features with shards. However, the rock magnetic signal also leaves fingerprints in rocks in a different way and with a completely different application, as we will see below.

---

[23] Geologically for a quick look at the sediment layers, here especially their lateral extent.

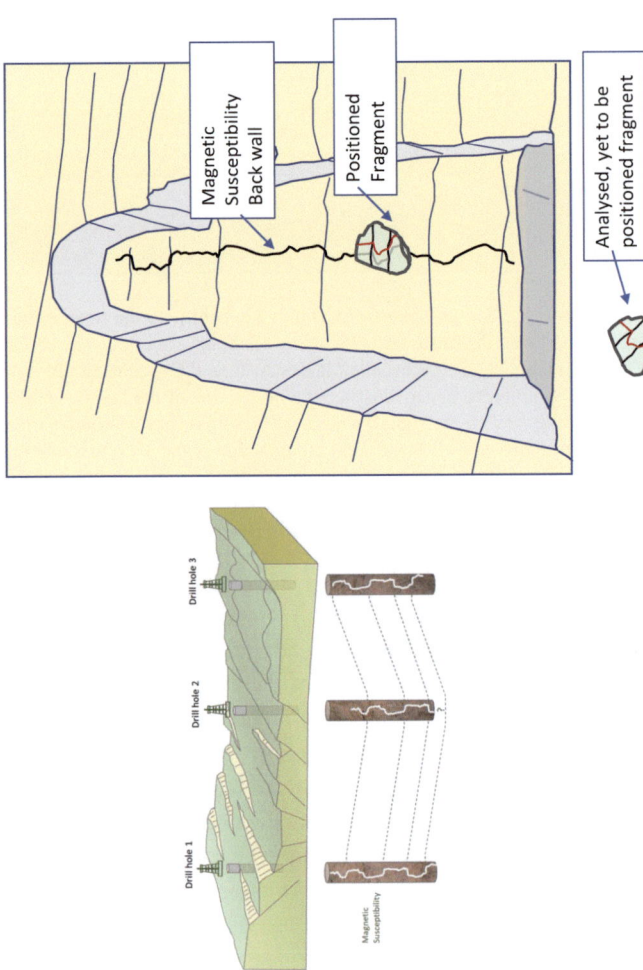

Fig. 5.20 The positioning of a fragment is based on a pattern comparison of magnetic susceptibility and is carried out in three principal steps. 1) A magnetic profile of the rock in place (here in the area of the back wall of the niche) is measured perpendicular to the geological stratification as a reference. 2) A magnetic profile perpendicular to the stratification of the fragment is measured. 3) The fragment is assigned to the corresponding place on the back wall based on its magnetic pattern (here susceptibility). As a rule, more parallel reference profiles are required than shown here, especially to adequately capture more complex geological situations. The geological layers in Bamiyan listed here as an example represent a simple geology due to their almost undisturbed, horizontally layered sediments, at least when viewed from a distance. Even with these sediments, lateral and vertical variation of the layers and thus the need for several magnetic profiles for reference can be expected in the centimeter range. The smaller the fragments to be assigned, the closer the measurement points of the magnetic susceptibility must be set to each other. The text discusses a sensible upper limit of resolution in geological scenarios. This once again underlines the necessity of conducting magnetic (any geoscientific) investigations against the background of geological relevance. The three drill holes in the schematic illustration of a regional depositional environment remind us of the corresponding procedure of a pattern matching in the correlation of distant drilling profiles. In such purely geological investigations, either one of the drilling profiles or, if available, cored rock of a drilling can be considered as a "fragment".

In the analysis of the magnetic properties of the rocks, we can go one step further—beyond the use of the general magnetic susceptibility. In other words, we have now reached the point where we open the black box of magnetic susceptibility a little further. So far that we now can actually look at individual grains and their properties. So far that we can focus on the properties of the often mentioned small, but "influential" group of ferromagnetic remanent magnetic minerals. Not only, after all, we want to answer geological questions. And this always includes the consideration of the entire rock.

What can we learn from a study specifically of remanent magnetic minerals in a rock? Do these minerals change over time—like iron rusts? Does the shape of the grains have a significance? Questions that we now want to investigate.

Traces of the Environment Millions of Years Ago

If we want to approach more geological questions and their answers through the rock magnetic properties of the rocks, already the headline "Traces of the Environment Millions of Years Ago" may seem like a riddle without geological background. What is meant by environment in the geological past? For this reason, we will first allow ourselves another brief consideration of geology, specifically sedimentology, because without such considerations it will again hardly be possible to understand the benefit of special rock magnetic analyses.

From the multitude of possible geological processes, we want to highlight two scenarios in the deposition of sediments:

1. Alignment of the grains and pore space during sedimentation
2. Contact with water and bacteria after deposition

In which direction did the river flow?

When sediment load is transported in a flowing body of water (a river, at the seashore, etc.), it is easy to understand that only a sufficiently strong current can transport a particle (for example, a quartz grain) of a certain size, a certain weight, or a certain mass. A gentle breeze rustles through the leaves of the trees, a hurricane may uproot the entire tree.

In relation to the size of the quartz grains in a river, this means that larger grains will settle at a certain flow speed, while smaller particles are still transported further. This applies in the first approximation.[24] Basically, our knowledge applies not only to

---

[24] Those who want to find out more should look under the keyword "Hjulström curve" in sedimentological literature to understand the more precise relationship between transport, deposition, and erosion, e.g., in Hjulström, F. (1935). Studies of the morphological activity of rivers as illustrated by the River Fyris. *Bull Geol Inst Upsala* 25:221–527, and many subsequent studies.

the different areas of rivers but also to all other depositional environments of the earth's surface where rock material is transported by water or wind. Dunes in a sand desert "migrate" in this way.

Not unexpectedly, on closer inspection, one must consider how the individual particles are transported. Let's stick with a river as an example. The finest particles float in the water, many others are either transported rolling, hopping, or possibly grinding on the river bed. What happens to a particle depends on the grain size and grain shape in relation to the flow speed of the water. Interestingly, water (as a mass, so to speak) also moves differently depending on the flow speed, sometimes in laminar "layers", sometimes flowing turbulently with lots of vortices. At this point, it is again not possible to go into the various environmentally relevant scenarios under which the flow speeds of waters or generally the deposition conditions of sediments in a geological space may change.

Two examples may serve as an indication that a closer examination of a sedimentary rock will allow conclusions to be drawn about the environmental conditions of its formation.

First, we imagine a very fine-grained sediment, the grains of which are just barely visible without a magnifying glass.[25] Either due to slight color differences of the grains or a slight change in the grain size of the components, we identify a stratification in the millimeter range. Such flat, undisturbed, parallel laminae[26] are intuitively not suspected as deposits of a high-energy milieu, for example at the bed of a raging river. Rather, we conclude a very quiet deposition milieu, for example in a lake without notable currents, in which the deposited material has primarily sunk slowly through the water column.

However, there is another intriguing thought in the context of our magnetic considerations: let us imagine the individual grains, no matter how small they may be, as small ellipsoids (or two-dimensionally considered as ellipses and not as circles) and how these elongated grains sink through the water column. When these particles come to rest on the lake floor, we will expect, as long as the grains do not touch, that the grains will lie with their long axes parallel to the lake floor (Fig. 3.5)—just like a grain of rice falls over when we try to balance it on its tip on the kitchen table (who wouldn't). If we always use a new grain of rice in the attempt to balance a grain of rice, the table will end up covered with grains of rice, whose long axes, parallel to the table top, otherwise point in all directions completely unordered. There is no force, so no reason, for a uniform alignment of the grains along their long axes on the table top (Fig. 5.21; see also Fig. 3.15 for a three-dimensional representation of ellipsoid as well as order and disorder).

---

[25] "How big is that?" is a certainly justified question. We ignore the individually different visual acuity and give here 0.1 mm as a reference value for an average "unarmed" eye.

[26] Geologically for "fine stratification".

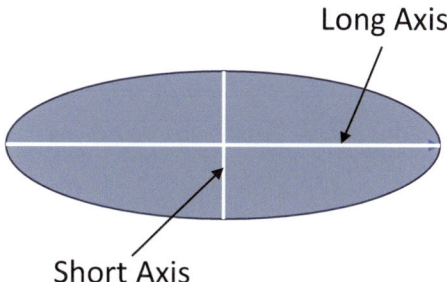

Fig. 5.21 Two-dimensional, schematic representation of an elongated sediment grain as an ellipse. An ellipse is characterized by the ratio of longer to shorter axis. If a sediment grain deviates from the ideal shape of a sphere as shown here, the orientation of its long and short axis will be relatable to the horizontal (as in the text example: a table top or a sediment surface) assuming no further forces

As a second example, we consider a rock with significantly larger components—the largest in the centimeter range—but with an equally large number of smaller, possibly angular, broken grains. For the depositional environment, we need a strong transport medium, due to the poor grain size sorting an unsteady transport medium, and due to the broken, poorly rounded components a geologically short transport path (or a short time) from erosion to deposition. Varying flow velocities in a river may be considered, or the transport of material by the grinding movement of glacial ice could provide an explanation.

In terms of our magnetic considerations, another essential difference will immediately become apparent: The grains—regardless of their size—do not settle out of the vertical water column,[27] but they are transported parallel to the Earth's surface. So there is a force that will directly affect the alignment of the long axes of our ellipsoids. As if we were now gently, but constantly and parallel to the table top, blowing on our unsorted rice grains. For sedimentary processes, we have already stated that grains may be transported bouncing, jumping or rolling—somewhat less gently than by blowing on a table top.

The essential point, however, remains: There is a force parallel to the Earth's surface (blowing, water, ice, wind) with the potential to align grains along their direction of transport (Fig. 5.22 and 5.23).

In other words, a person trained in sedimentology will be able to recognize from features such as size (better grain size distribution; Fig. 3.9), shape (predominant degree

---

[27] In Chap. 3 we considered this as a suitable scenario to store the Earth's magnetic field in sediments in the sense of a detrital remanent magnetization. However, in our consideration here, an aligning force of the Earth's magnetic field plays no role—the mechanical forces of a transporting medium are much stronger, and moreover, we are considering here commonly "non" magnetic grains (Sect. 3.1), for example quartz grains.

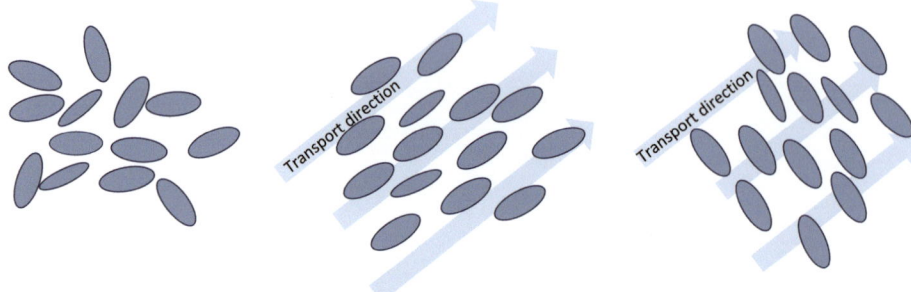

Fig. 5.22 Three different cases of alignment of sediment grains, each with a view from above onto the Earth's surface. On the left, the grains have remained without any order without the aligning force of a transport medium (water, wind) (for example, sinking in still water). In the middle and on the right, two basic possibilities of alignment of elongated grains are shown schematically. With very slow (example: muddy mass slides down a slightly inclined slope; Fig. 5.23), but also very fast transport speeds of the grains (example: center of a mountain river), the grains align with their long axes perpendicular to the direction of transport. In the middle, a typical alignment with the long axes parallel to the direction of transport is shown at medium speeds (example: current at the sea floor). Geometrically, elongated grains thus usually take one of two positions rotated by 90° when deposited (see also Fig. 5.23)

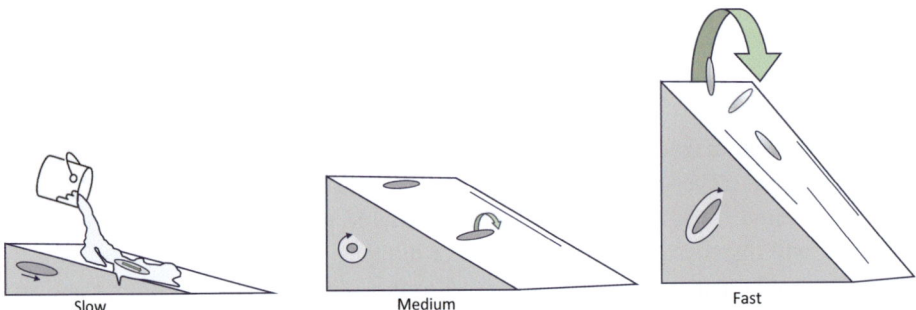

Fig. 5.23 Longitudinal alignment and transverse alignment of the grains depending on the speed of the transporting medium. Slow and fast transport both result in an alignment with the long axis of an elongated grain parallel to the direction of transport. To illustrate slow transport, imagine the creeping of a viscous mass on a flat inclined plane and for fast transport a bouncing and jumping, elongated pebble on a steep slope. At a lower inclination (medium transport speed), the same pebble would roll down like a roller

of rounding), arrangement of the grains to each other (type of packing) whether a rock is a typical lake bottom sediment, originally transported by the ice of a glacier or aeolian with the dunes of a sand desert. We cannot go into all possible sediment types and their formation processes in the context of this book. But we can understand a particular property of such sediment structures, their practical use and especially the possibility of making this visible magnetically.

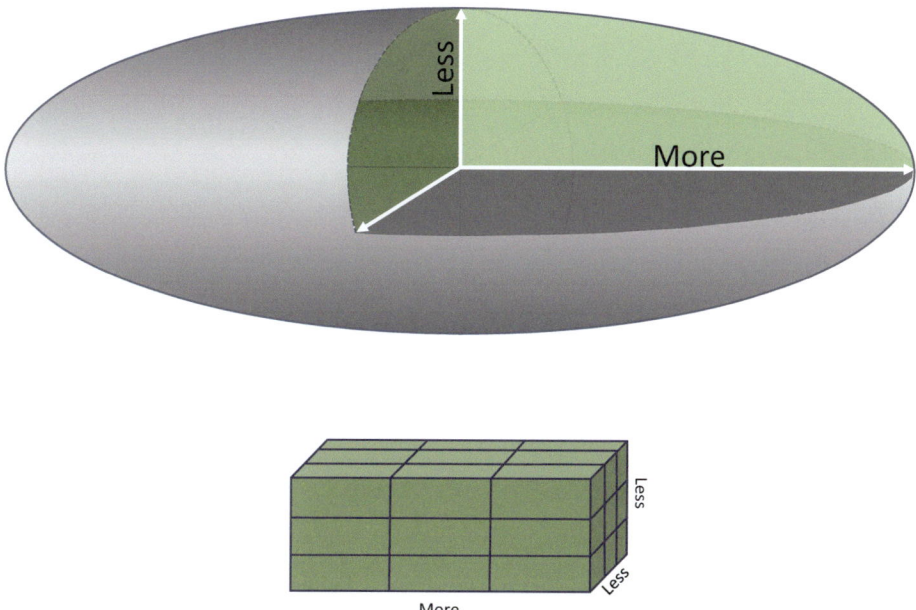

Fig. 5.24 Anisotropy often manifests itself in minerals and rocks solely through the shape of the objects. The magnetic susceptibility, for example, is greater when more of the same material (mineral) is present. A grain rounded to an ellipsoid has a higher magnetic susceptibility along the long axis than along the shorter axes. This applies in a simple way only when the internal structure of the grain in question is homogeneous, i.e., no anisotropy due to the crystal lattice is to be expected. A crystal lattice that is the same in every spatial direction can be imagined as stacked cubes (Fig. 5.25)—an anisotropic lattice is not the same in every spatial direction, for example in the sense of stacked shoe boxes (Fig. 5.25 right). Now one should not imagine the effect of shape anisotropy on the measurement of magnetic susceptibility too dramatically. The measurement in one direction of the mineral grain may only be about 1% different than the measurement at right angles to it. As long as this seemingly small difference—the anisotropy—can be reliably detected by our measuring device, this value can also be subjected to analysis and interpretation.

When we analyse a sedimentary rock, whether in a classical geological way or as here with rock magnetic methods, we naturally only look at those grains which actually settled under the conditions summarized above. How the grains settled—their arrangement to each other depending on shape and size—we then call sediment structure. And we want to make this sediment structure visible with a magnetic measurement. For this, we turn again to the already known magnetic susceptibility, but in a somewhat more specialized view, because so far we have only considered this characteristic as a lump sum value (as a pattern) along a rock profile. Fig. 5.24 explains to us that upon closer inspection, that is, when measuring the susceptibility of the same sample in different orientations, different values could be determined. Here lies the possibility, simply put, to characterise an elongated grain (or several similarly oriented in a rock sample) magnetically—ideally to characterise a sediment structure magnetically. As a rule, we assume in

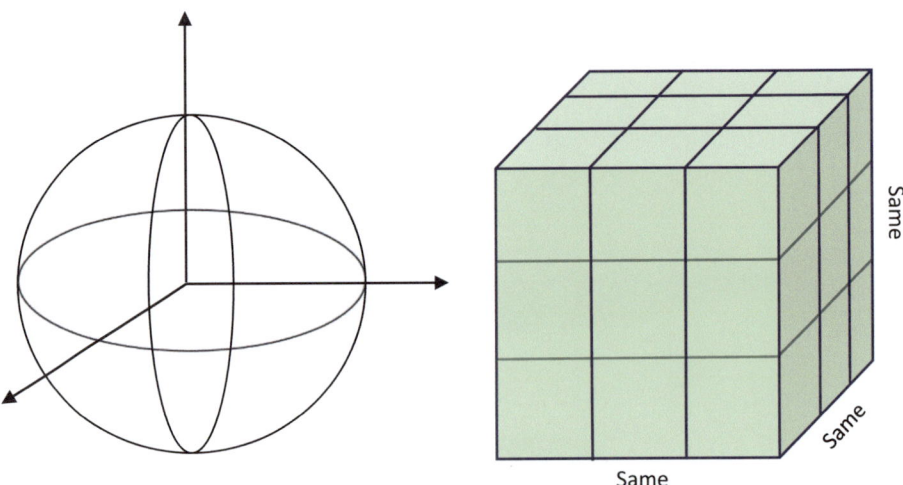

Fig. 5.25 Side by side of a sphere and stacked cubes to illustrate isotropy. The same substance, the same quantity, and accordingly the same strength of magnetic susceptibility are present in three orthogonal spatial directions. Note: Even a mineral grain that is perfectly spherical in its external form can exhibit anisotropy when measuring magnetic susceptibility if its internal crystal lattice is constructed differently in the three spatial directions (Fig. 5.24)

this approach that the measurements of susceptibility are a function of the external shape of a mineral grain (for example, the quartz grains in a sandstone).

So now we finally come to the point: If we want to magnetically capture the orientation of the grains in a sedimentary rock, we have to measure a magnetic anisotropy (Fig. 5.24)—a property, therefore, which gives a different value for the same type of measurement in one direction of the sample than perpendicular to it. In the simplest case, we again use the magnetic susceptibility, because this magnetic characteristic is "quantity-dependent" and thus higher along a long axis, since more material is captured in relation to the shorter axis perpendicular to it (Fig. 5.24). In a book, which is intended to contribute to the basic understanding of magnetic operation, it is not appropriate or necessary to go into the special procedures of the measurements for magnetic anisotropy. It is certainly worth mentioning that the respective rock samples must again be taken oriented from the rock association, as already explained in Chap. 4. A magnetic anisotropy measurement can then be carried out in a suitable measuring device in strict relation to the shape and orientation of a sample. Certainly, for all investigations, several samples will be taken, often along a geological profile and thus over a certain period in earth history, in order to be able to trace, for example, the changes in the flow direction of a long-dried-up river. As a rule, a large amount of data will be generated in this way. For help in analyzing the data, so-called stereographic projections are often used. With Fig. 5.26 we briefly devote ourselves to the description of such methods, in order not to completely lose sight of any direct reference to hard data in all generalizations. It should be noted

5.2 Applications of Rock Magnetism in the Geosciences

that in measurements of the anisotropy of magnetic susceptibility, basically the orientation of the long and short axes of an ellipsoid is measured. We can consider one axis as a linear, as in Fig. 5.26, and represent it in its geographical orientation.

For a practical example of the application of such an anisotropy measurement, the essential steps for determining the flow direction of a river in the geological past are explained in Fig. 5.27 and 5.28.

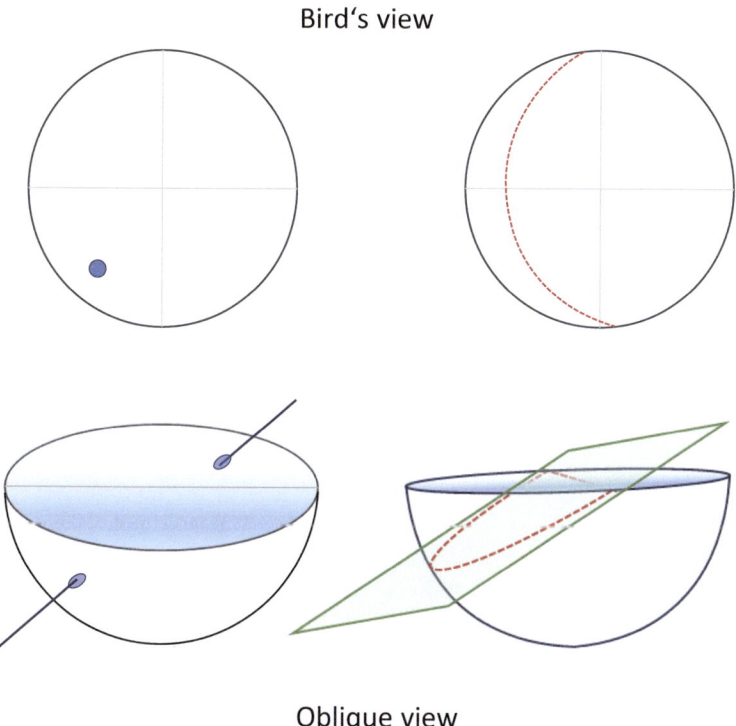

Bird's view

Oblique view

Fig. 5.26 The orientation of surfaces or lines on the earth's surface can be graphically represented using stereographic projections. Imagine a surface in its orientation relative to the entire globe and thus the network of longitudes and latitudes (Fig. 2.7). If our surface is the side of a book or our smartphone, we can mentally extend this surface to the size of the earth (or shrink the earth to the size of the book page …) and move it to the center of the earth without changing the orientation itself (bottom right). If we now look from above into the (inner) lower half of the earth—like into a bowl—we see the trace in which our book page cuts the earth's surface (dashed line). Top right is a stereographic representation of the orientation of our surface (dashed line) in this sense. If we had held the book more upright, the dashed line (top right) would have been less curved and moved more towards the center of the circle. If we mark four points of the circumference in the illustrations with north, south, east, and west, we have a direct reference to the geographical system of longitudes and latitudes. In a similar way, a linear (for example a pen or the axis of an ellipsoid as in Fig. 5.24) can also be represented in a stereographic projection (bottom left). The point at the top left represents the inclination angle and orientation of the "pen" in this way. When evaluating magnetic results for anisotropy, one will often look at such points and point clouds (several samples) on circles (stereographic projections)—this type of graphical representation is often clearer and simpler than drawing quantities of elongated mineral grains

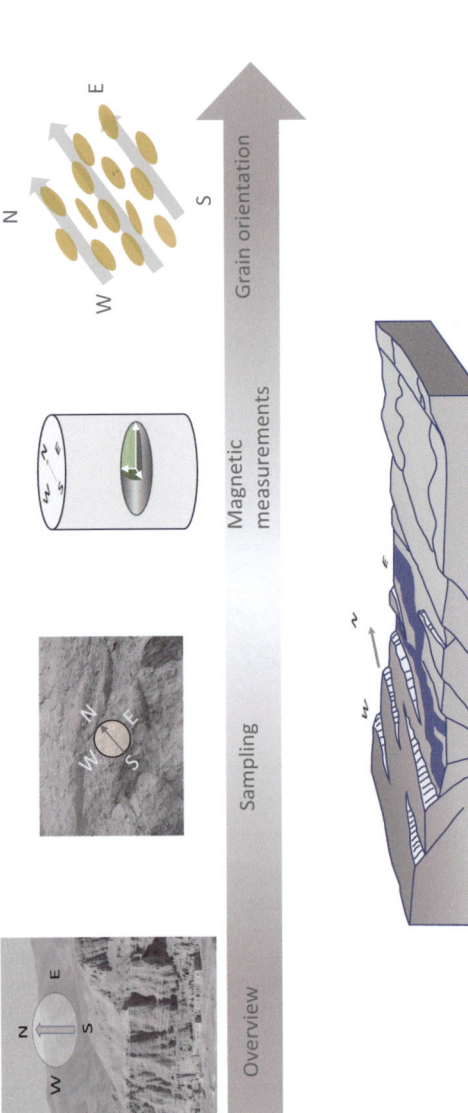

Fig. 5.27 Highly schematic summary of the essential steps to determine the flow direction of a river from a sediment sample. (1) We examine a layer at a cliff (Overview). The geological study suggests a typical deposition of the components in a river bed. (2) We drill a number of sample cylinders so typical for magnetic investigations and determine the orientation of these samples in the geographical system of north, east, south, and west (Sampling). (3) We measure the magnetic susceptibility of the samples in different directions relative to the sample. In this way, we determine the anisotropy of susceptibility or, in other words, the direction of the long axis of the sediment grains in relation to the sample cylinder (Magnetic measurement). (4) Since we know the orientation of the sample at the time of collection, we can also determine the orientation of the long axes of the grains in relation to (geographically) north. The measurement is based on the assumption that all (elongated or also flat, i.e., not spherical) components of the rock sample were aligned in the same way during the original deposition by some transport medium (Fig. 5.22). For the evaluation, some calculations, so-called coordinate transformations, are usually necessary to correctly represent the orientation in the geographical system (Fig. 5.26). It will rarely be possible to take samples exactly horizontally and in a precise north-south orientation. Perhaps the layer from which the sample originates was also tectonically displaced after the layer had been deposited. Perhaps the entire deposition space has rotated due to plate tectonic processes since the deposition. All of this must be corrected for the final result (5)(river/direction in the past), but is irrelevant for the principle of the procedure. The grains remain unchanged in their position relative to each other, through the coordinate transformation only the space is rotated around the grains until space and grains can be viewed again in their original position relative to each other (Fig. 5.28)

5.2 Applications of Rock Magnetism in the Geosciences

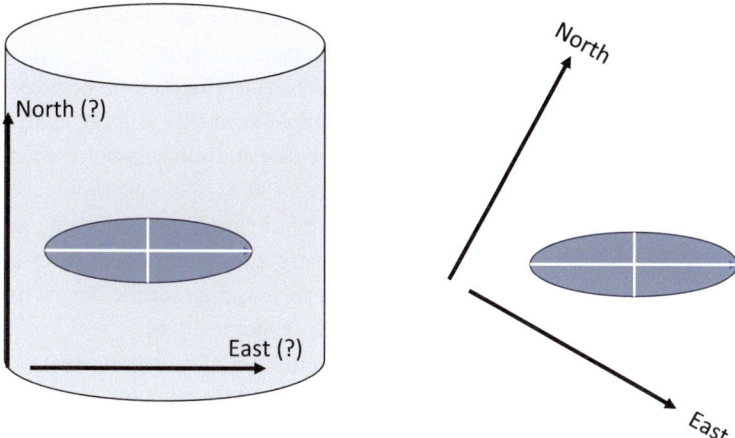

Fig. 5.28 Highly schematic and two-dimensional illustration of a coordinate transformation. When the orientation of the long axes of sediment grains (or corresponding values of the anisotropy of magnetic susceptibility) is determined, this is done in a first step in relation to the sides (the coordinate system) of the sample body (left). This is how the sample was drilled, and, more importantly, this is how the sample was inserted and measured in the measuring device. If the sides of the sample cylinder were conveniently oriented in the north/east direction during sampling, the orientation of the long axis of the grain can be read directly (east-west). In most cases, the sample cylinder will not have this ideal geographical orientation. The coordinate system must therefore be rotated until its axis is parallel to the geographical directions. Only then can the true orientation of the grains be read—in our example northeast-southwest. Often several such steps must be taken until the desired direction is determined. In geological practice, coordinate transformations are three-dimensional, so there is just one more axis to correct. The principle remains the same; however, the order in which the corrections of the individual axes are made must be observed. To illustrate, place a book on the table, turn it 90°clockwise, stand it upright and turn it another 90° clockwise. If the order of actions is reversed (for example 90°, 90°, upright), the book will end up in a different position despite the same starting position. Geologically speaking, this means, for example, correct the angle of the layer inclination first in case of a tilting of the layer and then turn to the north direction—not the other way around—etc.

We have considered anisotropy here using the example of magnetic susceptibility. In our considerations, we have assumed that the anisotropy results from the elongation of individual grains. Thus, we assume that the internal structure (crystal lattice; Fig. 3.1) is not a factor in determining this magnetic anisotropy and that only the different amounts of the same material are the decisive point.

Of course, there are minerals, also from the series of "strong and often dominant" remanent magnetic minerals, which already show a strong magnetic anisotropy due to their crystal lattice. A spherical particle (Fig. 5.25) of the mineral pyrrhotite (magnetic pyrite) shows this property, for example. A corresponding entry is therefore present on the data sheet of the diagnostic properties for this mineral (Fig. 5.2) and could be used in other investigations together with further magnetic criteria for the decryption of the rock magnetic inventory of this sample.

Such special magnetic investigations are not our topic in this book. At this point, we just want to use this small loop in our considerations to once again highlight the importance of integrated and coordinated measurements and analyses in geology. In other words: A "thoughtless" measurement of a rock sample is usually without analytical value if a single measurement value, here the magnetic susceptibility, is not embedded in an overall context for the sample.

In the same way, we have already postulated that a single measurement value of the blanket susceptibility or, as in Chap. 4, a value measured without context of a magnetic direction in a sample can lead to no usable result, for example for the diagnosis of a rock type or the state of the earth's magnetic field in the geological past.

In our example for determining the former direction of a bygone river, we need a sequence of additional findings to answer such questions, for example along the lines:

- It is a sediment (i.e., geological address).
- Elongated grains consist of an (magnetic) isotropic material (i.e., a sandstone mainly made of quartz grains for example).
- The tectonic status of sediment layers can be determined (coordinate transformation).
- The anisotropy of magnetic susceptibility can be measured with sufficient accuracy—such measurements can only be carried out with more complex equipment in a rock magnetic laboratory and not with portable measuring devices, as is sufficient for simple layer correlations (example Buddha).

The points listed are not a disadvantage of the applied magnetic methodology, but it is simply good practice not to draw pseudo-insights from isolated measurement values without context. In the following third example of the application of rock magnetic measurements, we will be a bit more daring and consider what traces water leaves in rocks through the subsequent alteration after deposition. Bacteria, really?

Magnetic minerals, fluids and bacteria

Don't worry, the following brief remarks are hopefully easy to bear even for pandemic-tested characters. Bacteria[28] are a quite ubiquitous component in sediments of the upper lithosphere and have been widespread worldwide since geological times—more on this at the end of the chapter. First, we consider a crucial prerequisite for the existence and spread of fluids such as water, oil or gas as well as organic components in rocks: the pore space of a rock.

As in other parts of this book, we use the term "pore space" primarily in a plakative way and not in a strict sense of a scientific definition. By pore space, we mean the

---

[28] No viruses!

5.2 Applications of Rock Magnetism in the Geosciences

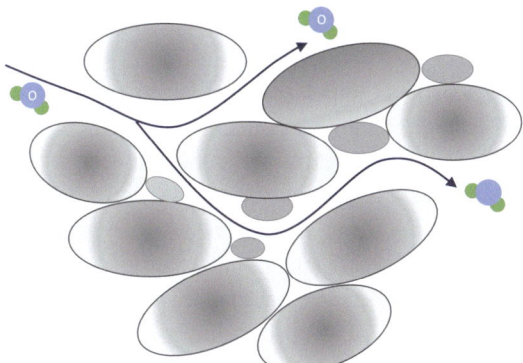

Fig. 5.29 As long as the molecules of a fluid, here indicated is water (H$_2$O), can still pass through the pores, such fluids can circulate in rocks. As long as fluids circulate in rocks or simply flow through, they will slowly and continuously change rock as well as unconsolidated sediment—to a different extent and variably fast depending on the composition of the rock and the type of fluid. The paint on a garden fence weathers faster in salty sea air than in the mountains far away from acid rain

space and passageways between the minerals/components of a rock, in this sense: With sediments, as always the exemplary rock group of our choice, the term is almost self-explanatory. Many sediments consist of individual, "collected components". So, as we have discussed the orientation of elongated grains in the previous section, it is easy to see that a free space remains between stacked grains (see also Fig. 3.5 and 3.6).

Finally, grains are irregularly shaped, and one does not fit into the other—Tetris with rock grains hardly works without leaving gaps. Certainly, depending on the shape and size of the pebbles and the way they are stacked, there are different effective packings—effective in the sense of minimizing the remaining gaps between the pebbles. In nature, rarely only equidimensional components/minerals are de posited—increasingly smaller grains fill the space between the larger pebbles. This makes the pores increasingly smaller or clogged.

Now, (ground) water, for example, or gas can only circulate in the sediment layers of the earth's crust[29], until even the finest pores are literally "cemented" or at least the connection between the pore spaces is blocked (Fig. 5.29). The cementation progresses, for example, through calcareous or silicate precipitations from the circulating water, just as lime in the household slowly reduces the cross-section of a water pipe.

Let's imagine somewhat simplified that water (H$_2$O) is made up of molecules like small balls—two hydrogen balls and one oxygen ball, whose respective and thus

---

[29] Circulate in the sense of "moving". Hydrocarbons, such as gas or oil, migrate underground from the place of their generation from organic components until they accumulate in a geological "trap" and become a deposit. Our groundwater is also known to move over often large distances deep in the earth.

composite total molecule size is fixed in nature. Very small by human standards, but still measurable and nameable.

Simplifying as always, what is important now is again dimensions: As soon as these tiny "balls" no longer fit through the (even smaller) pores, the circulation of the water is stopped. Components dissolved in water (we have cited lime as an example) or simply carried along are naturally also stopped at too narrow pores depending on their respective size. This is called filtering. If not mentioned yet: Many of the magnetic minerals in rocks are also tiny and can be carried along in the water for a correspondingly long time.

Water molecules have a slightly different size than some oils or gases. When which fluid can no longer circulate in sediments accordingly has different limits. Depending on the fineness of the deposited sediments, the depth to which fluids circulate varies; often the flow occurs in discrete layers with better permeability compared to the adjacent layers. Circulation of fluids hundreds of meters deep in the earth's crust is no geological surprise.

Often, already solidified rocks break due to tectonic processes deep underground, and often this is expressed in finest hairline cracks or even clear fractures of the rocks. Instead of through pore spaces, fluids then find their way along these fracture planes.

Here comes our crucial point: As long as water/other fluids can circulate through the rock, be it in the pore space or along discrete fractures, these fluids interact with the "rock". Interact in the sense of: altering the mineral content and possibly even the cohesion, i.e. the structure of the rock—somewhat comparable to when water runs over metal for a long time and rust increasingly forms.

To this, another crucial point: As for interaction of rock with circulating fluids, the group of remanent magnetic minerals (yes, this small illustrious group of ferromagnetics) are among the "most sensitive" minerals. The reason lies, for example, in the small size—small particles dissolve faster than larger ones—but also in their structure.

After oxygen, iron is also, so to speak, a sought-after energy donor. Also in rocks. At least for bacteria. "Dissolution" of at least individual components can therefore be understood in rocks as being "nibbled" by bacteria. Again, it should be mentioned that we cannot go into details due to the complexity of the topic. But as always, we want to understand the principle. For this, we imagine the seabed.

Muddy material has been delivered and deposited for hundreds of thousands of years. By now, a sediment thickness of 100 m and more has built up. With minerals such as quartz, mica or even finest ferromagnetic minerals, organic material will also always land on the seabed: eroded plant material, microorganisms from the water column, in some cases also the particulate remains of other organisms from the water column—the list is long and diverse.

A large part of the bacterial activity in the growing sediment column takes place in the more complex context of degradation of this organic material. Depending on the rate at which new sediment material accumulates, and depending on how much and what type of organic material is delivered with it, the degradation, the decomposition of this material follows strict chemical and physical rules.

5.2 Applications of Rock Magnetism in the Geosciences

As one simple rule, the degradation of organic material will continuously advance with increasing depth[30]. One of the end products is the formation of methane gas at depths. Bacteria need energy to break down the organic material, and a popular energy source is oxygen. When all the free oxygen is consumed with increasing depth, compounds of oxygen with other elements are used. Magnetic minerals are among the more popular energy carriers already during the early stages of degradation (Fig. 5.30), sometimes just a few centimeters below the surface (of the seabed in our example). Depending on the depth or, more accurately, the corresponding stage in the degradation of organic material with depth, you will find different types of bacteria, each adapted to the requirements of a specific degradation level, as if craftsmen in house construction were not only distinguished by their respective expertise from excavation to painting, but also by their preferred food.

Beware, with the next consideration we finally get to the point. What does all this have to do with rock magnetic studies? A lot in some cases, very little in others, would be one of the wise answers at this point.

If nothing changes during sedimentation (as in our example onto the seabed) over time, i.e. the amount of particles delivered as well as the type and amount of embedded organic material always remain the same[31], effectively little noteworthy happens for magnetic signals in these layers. When magnetite is bacterially broken down into its components iron and oxygen, part of the iron, for example, moves upwards[32] towards the seabed surface, and as soon as it encounters free oxygen, new magnetite precipitates (sometimes there are also special bacteria to accomplish this). This is often not a long journey and is certainly completed in the uppermost few centimeters of the sediment column. After that, in the ideal case—in terms of the magnetic signal—everything

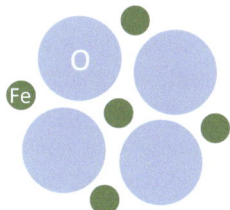

Fig. 5.30 Many of the remanent magnetic minerals consist of a combination of iron (Fe) and oxygen (O). When there is no more free oxygen available as a preferred energy source in rocks (suboxic area in sediments), some specialized bacteria in the depth of the earth's crust also break up combinations of Fe and O to get to "food". How quickly an attacked particle is completely "dissolved" is a question of the surface area of the particle. In other words, the larger a particle is, the longer it takes, even if more bacteria are working on this surface at the same time.

---

[30] Or in other words: more material in the overburden and over a longer period of time.

[31] Steady-state conditions in the scientific sense.

[32] The movement follows a chemical gradient, not gravity.

is as before. Depending on how quickly the process of dissolution and re-precipitation takes place, perhaps only a small part of the magnetite minerals present in the sediment is affected anyway. The dissolution and precipitation front, so to speak, has passed too quickly to catch all the minerals. The rest remains unchanged.

If even one of the parameters changes during sedimentation, for example if suddenly more sediment material is delivered to the seabed, the boundaries of the degradation process shift spontaneously. For example, the boundary at which free oxygen is still available moves spontaneously to a different depth. This can easily happen in nature. Imagine the slow and steady sedimentation on the seabed—and suddenly a huge package of sand and mud deposited further up towards the coast slips off and pours out like an avalanche over this area. Such a turbidite almost spontaneously raises the seabed at the point of its deposition by meters. At depth, this means, from a chemical point of view, a spontaneous shift of boundaries, which must then be adapted to the new depth. The same applies, of course, to all the highly specialized bacteria and their degradation activity.

From a magnetic perspective, this means that magnetic minerals, which otherwise would have been dissolved at a certain point in the sediment column or precipitated elsewhere, do not come into play or, in other words, that the magnetic signal, which we correlated at the beginning of this chapter with the lithological boundaries of the sediment column, will show secondary "bumps and dips" in the course of the measured magnetic characteristics (Fig. 5.31).

The task then is to interpret the cause of the unusual "bumps and dips" and trace them back to the underlying geological processes. Certainly an exciting part of any investigation. From a magnetic perspective, one will not only use one characteristic, for example the much-cited magnetic susceptibility, for such investigations. One will try to characterize in particular the illustrious group of remanent magnetic minerals more precisely. fFor example to determine whether magnetite, if present, shows a grain size distribution, which suggests several formation processes. Or whether in addition to "pure" magnetite, different types of magnetite are present, which are oxidized, in a vivid sense rusted. In relation to Fig. 5.2 and 5.3, a variety of specialized measuring devices and methods are used to decipher the rock magnetic signal of a sample. These special laboratory studies are used, in other words, to open and read out the black box of the general magnetic susceptibility.

5.2 Applications of Rock Magnetism in the Geosciences

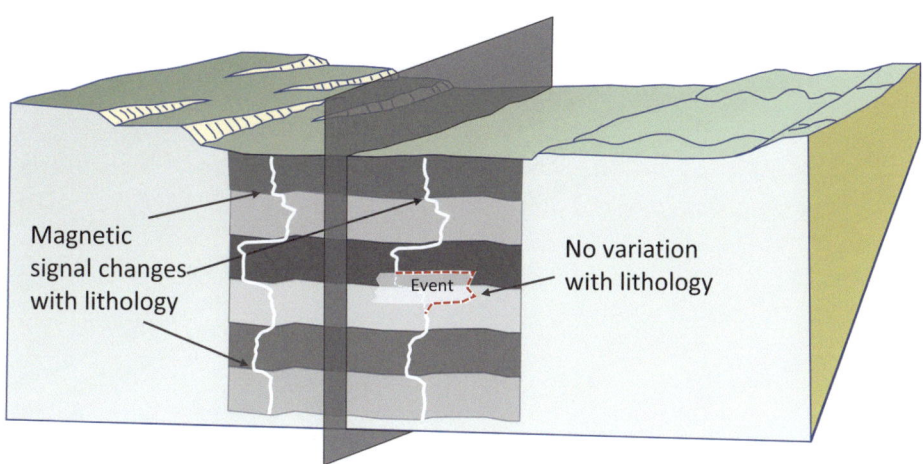

Fig. 5.31 View of a schematic sequence of nearly horizontally layered rocks (different shades of gray). A scale is deliberately not given. The vertical depth could range from meters to hundreds of meters depending on the actual geological situation. As already explained, we usually expect a variation of the magnetic signal with changes in the rock composition (see example Bamiyan). Most clearly at layer boundaries. For further graphic illustration, a fictitious "vertical wall" (gray area) separates the left geological scenario from the right side of the illustration. On the left, the white curve represents the expected change of the magnetic signal with the lithology. In the simplest case, the white curve represents, for example, measurements of the magnetic susceptibility of the layers. In a slight modification of this idea, it helps for the next thought process to consider the white curve as the quantity, type, and composition of the ferromagnetic minerals in these layers. The text explains how these minerals dominate the magnetic signal, but also react most sensitively to any kind of change in the sediments after deposition. On the right, such changes are symbolized by the deviating magnetic signal in the area of the "event" marker. An example of such an event, where the magnetic signal changes subsequently, is discussed in the text in connection with the degradation of organic material in the sediments. However, many other scenarios are conceivable, ranging from the migration of groundwater or oil in individual layers to the thermal heating of layers near hot magma deep underground. Change of the magnetic signal in this sense means that, for example, a part of the originally detritally deposited remanent magnetic minerals are dissolved in the sediment by chemical-physical processes (in the text we use the more vivid, but quite typical example of bacteria attacking iron oxide and magnetite in the pore spaces of the rocks) and are no longer present or replaced by another mineral. In the text, we refer to the magnetic susceptibility as a black box, into which one must look to reveal the individual contributing minerals. Special rock magnetic studies in the laboratory make it possible to identify, particularly ferromagnetic, minerals based on their specific characteristics and properties in rock samples and to interpret them accordingly with regard to "events" in the history of the rocks. The mostly over 99% of the "non-magnetic" minerals such as quartz or feldspars in a sedimentary rock react less sensitively to chemical-physical-bacterial changes and are therefore hardly the subject of such environmental magnetic investigations.

Conclusion and Outlook   6

Are paleo- and rock magnetism now "niche" sciences? Judging by the application in Chap. 5 to the niche in the Hindu Kush, certainly. If one disregards this attempt at wordplay, not so much.

Today's understanding of our Earth is based on the ever-advancing, ever more refined research in a multitude of geoscientific disciplines. In the same way, individual areas of paleo- and rock magnetism are also continuously developing. A look at recent research contributions, including presentations in the sessions of international geoscientific conferences, shows a focus on some of the topics of this book over the years: dating, environmental analyses, plate tectonics, physical basics and evaluation methods in rock magnetism, and high-resolution, ideal recorders of magnetization in the rocks of the Earth's crust. Especially for the analysis of the climate in the Earth's past or, of course, for a better understanding of the other previously mentioned focal points, "ideal" rocks of the Earth's crust are sought-after research objects among paleomagnetists.

In this book, we have only presented a small selection of possible applications for the magnetism of rocks, especially to understand the underlying principle of paleo- and rock magnetic methodologies.

On this basis, we can now conclude with a brief outlook on the more complex integration of scientific investigations based on paleo- and rock magnetic methods into the contexts of the Earth system.

For the demonstration of a promising recorder of magnetic (environmental) information, let's take a globally widespread "dusty" sediment: loess[1]. This aeolian sediment has

---

[1] A wind-transported, clastic sediment with a typical grain size distribution below what we would commonly call sand. These silts, often of a yellowish-brown color, have been blown out of their original location like dust, transported, and later accumulated in hundreds of meters thick deposits. The extensive loess plateau in northern-central China is an example. The (stable) geological

© The Author(s), under exclusive license to Springer-Verlag GmbH, DE, part of Springer Nature 2025
M. Urbat, *Magnetism of Rocks*, https://doi.org/10.1007/978-3-662-70428-8_6

been the subject of magnetic research for good reason for several decades. In Chap. 2 we considered general criteria for particularly suitable rocks for magnetic investigations. Loess has further advantages in this respect, especially for understanding climatic variations in the recent Earth history:

The deposits occur consistently year after year in "sufficient" quantity. Let's say for a few hundred thousand years. The type and amount of loess during this time are varied on a higher level by the change from glacially cold (glacial) to warmer interglacial periods of the climate. Just as it rains more in some summers and less in others in everyday life, the amount, composition, and grain size distribution of the loess accumulation also vary. Even in a stable situation, where the source, transport, and deposit area of the material remain essentially the same, small variations are enough for changes in the deposits - perhaps a slightly lower wind strength. In the warmer phases of the glacial recent Earth history, soil formation[2] also begins in the upper layers of the deposited loess.

A loess profile that is several hundred meters thick today will accordingly contain several such, now fossil soil horizons as witnesses of the climatically warmer phases. We already suspect that soil formation will also affect the magnetic minerals of such layers (see Sect. 5.2, Magnetic Minerals, Fluids, and Bacteria). Finest magnetic minerals, such as magnetite, which are blown in, are "nibbled" during soil formation by bacterial activity and even additional magnetite is precipitated elsewhere. Imagine a profile of magnetic susceptibility (similar to Fig. 5.19 in Chap. 5) will show an increased deflection for each soil.

Now, the causes of the natural fluctuations of the paleo-climate also in the Earth's past and the possibility of reconstruction are complex. We want to highlight two factors as examples:

Orbital parameters of the Earth have a significant influence. On its elliptical path, the Earth moves closer and further away from the sun, and with its (more or less wobbling) axis, it is more or less inclined towards the sun (see Chap. 2). Such fluctuations occur cyclically with different frequencies[3]. Better known in the geosciences as so-called

---

situation in the recent Earth history (Quaternary) with, among other things, the Gobi Desert as a source, glacial climatic variations, and monsoon dominated winds has created a high-resolution recorder of the paleoclimate with these loess deposits. There are other, comparable scenarios worldwide.

For a magnetic introduction, see, from the numerous more recent publications, for example: Evans, M.E. and Heller, F. (2001). Magnetism of loess/palaeosol sequences: recent developments. *Earth Science Reviews*, Volume 54, Issue 1–3, 129–144, Elsevier.

[2] Keyword pedogenesis. In the broader sense, the weathering of the upper layers of the loess results in fertile soil horizons, which are later covered with fresh loess under colder conditions.

[3] Similar to how repeating patterns within a pattern are created with a Spirograph. This is a "toy" invented about 180 years ago, in which gears are moved into each other with a drawing pen to draw mathematical patterns.

Milanković cycles[4]. Such cycles form an essential framework for climate modeling of Earth's history, for example, the waxing and waning of ice sheets.

A suitable proxy for climatic (cyclic) fluctuations in the recent Earth's past is the standardized and dated oxygen isotope curve[5]. Due to the sensitive reaction of the magnetic minerals to any kind of environmental conditions (see above for example soil formation in the relatively warm phases of the Ice Age), a good agreement of the susceptibility and other rock magnetic characteristics of the loess sequences and the standardized oxygen isotope curve can be demonstrated[6]. In this way, the magnetic signal of the rocks becomes a climate signal. Or rather, a proxy for the climate in Earth's past. Especially when the respective curves of the magnetic susceptibility are subjected to a so-called frequency analysis[7] and Milanković cycles can be demonstrated.

From what has been said here, it can be further concluded: the more precisely the rock magnetic signal of a loess sequence can be deciphered, the more precisely the causes of the respective changes in the environment can be deduced. The more precisely it can be determined what exactly is hidden in terms of minerals and possibly different generations of minerals in the black box of magnetic susceptibility (Chap. 5), the better the image of the rainfall, air temperatures, and wind direction will be reflected in the magnetic minerals of the rocks.

Certainly, one can also simply perform a magnetostratigraphic dating in loess sequences. The better we know the underlying rock magnetic signal, i.e., the composition of the minerals, the better we can assess the supposed recording of the Earth's

---

[4] Named after the Serbian mathematician and geoscientist Milutin Milanković, 1878–1958. Known cycles range from a length of days to the geologically more prominent cycles with frequencies of, for example, around 20,000 to 400,000 years.

[5] Isotopes of an atom, for example, of oxygen, differ by number of neutrons in the nucleus, and thus have a different atomic mass. For seawater, for example, one might specify an average, typical ratio of the stable oxygen isotopes of the water molecules. Water now evaporates, for example, in the course of the global water cycle over the sea or elsewhere. Water thus changes into another state of aggregation (vapor). In this process, the ratio of the stable isotopes can shift, as heavier or lighter "building blocks" can preferentially be incorporated into the newly emerging water vapor or can remain in the water. A driving factor, also in the atmospheric circulation for our example of evaporation, is the ambient temperature (of the air). A curve of the oxygen isotope ratio over thousands of years can accordingly be seen as an indicator of temperature fluctuations (in the broader climatic context also sea level fluctuations). Such curves of minimal fluctuations for Earth's history are often measurable on the calcareous shells (fossil) of marine organisms or also on ice cores.

[6] E.g. Liu, Q., Banerjee, S.K., Jackson, M., Deng, C., Pan, Y, and Zhu, R. (2005). Inter-profile correlation of the Chinese loess/paleosol sequences during Marine Oxygen Isotope Stage 5 and indications of pedogenesis. *Quaternary Science Reviews*, Vol. 24, Issue 1–2, 195–210, Elsevier.

[7] Simply put, one calculates the frequency with which higher and then lower values of susceptibility alternate over time. See above, for example, every 20,000 years etc.

magnetic field. For example, whether there was a delay in the recording (Chap. 3), because bacteria were involved.

It is clear - especially the ongoing development of high-resolution measuring devices on the physical basis of magnetic research as well as the further development of mathematical algorithms for analysis and especially pattern recognition in large data sets will remain a prominent topic in paleo- and rock magnetism.

MIX
Papier aus verantwortungsvollen Quellen
Paper from responsible sources
FSC® C105338

If you have any concerns about our products,
you can contact us on
ProductSafety@springernature.com

In case Publisher is established outside the EU,
the EU authorized representative is:
Springer Nature Customer Service Center GmbH
Europaplatz 3, 69115 Heidelberg, Germany

Printed by Libri Plureos GmbH
in Hamburg, Germany